Experimental Erosion

Xiangzhou Xu · Tongxin Zhu ·
Hongwu Zhang · Lu Gao

Experimental Erosion

Theory and Practice of Soil Conservation
Experiments

Xiangzhou Xu
School of Hydraulic Engineering
Dalian University of Technology
Dalian, China

Hongwu Zhang
State Key Laboratory of Hydroscience
and Engineering
Tsinghua University
Beijing, China

Tongxin Zhu
Department of Geography, Urban,
Environment and Sustainability Studies
University of Minnesota Duluth
Duluth, MN, USA

Lu Gao
School of Hydraulic Engineering
Dalian University of Technology
Dalian, China

The published book is sponsored by the Dalian Municipal Government.

ISBN 978-981-15-3803-2   ISBN 978-981-15-3801-8   (eBook)
https://doi.org/10.1007/978-981-15-3801-8

Jointly published with Science Press
The print edition is not for sale in China (Mainland). Customers from China (Mainland) please order the print book from: Science Press.

This Springer imprint is published by the registered company Springer Nature Singapore Pte Ltd.
The registered company address is: 152 Beach Road, #21-01/04 Gateway East, Singapore 189721, Singapore

Foreword by Guangqian Wang

Soil erosion is one of the major environmental problems facing the world which is reducing cultivated land areas, decreasing land productivity, increasing riverbed, and causing reservoir sedimentation. The Loess Plateau, where the case studies of this book take place, is among the most severely and adversely affected regions by soil erosion in the world. The study and control of soil erosion would deliver enormous environmental, social and economic benefits to the society.

The book provides a comprehensive overview of the theories and principles for soil conservation experiments. It also proposes a systematic and rigorous methodology for quantitatively observing and measuring erosion processes, including more than ten patented devices and technology invented by the authors of the book. For example, the easily assembled mobile laboratory makes it convenient to perform rainfall erosion experiments in the field; the patented Topography Meter enables to detect the dynamic changes of slope surface terrains during a rainfall experiment and thereby to calculate the volume of gully erosion and gravity erosion.

The most significant contribution of this book is the experimental studies on gravity erosion. Gravity erosion, e.g., landslide, avalanche, and mudslide, is typically investigated in the field after its occurrence, due to the difficulty and uncertainty in predicting the timing, frequency, magnitude, and specific location of its abrupt occurrence. There is a scarcity of gravity erosion process data. The conditions of a slope failure, e.g., soil moisture and shear strength, surveyed after its occurrence may be significantly different from those prior to and during its occurrence. It is difficult to know how much sediment produced by slope failures is transported during the current rainfall and how long it will take to transport the remaining debris in the subsequent rainfalls. The answers to those questions are of critical importance in assessing the contribution of gravity erosion to the sediment yield within a watershed. In order to address those important questions, numerous exploratory experiments have been conducted on the steep slopes both in the laboratory and in the field of the Loess Plateau by the authors and their students/research partners in the past two decades. With the state-of-the-art equipment invented by the authors of the book, both the occurrence of slope failures and the sediment the failures generated can be monitored during the experiments. Those experimental studies are concerned with the behavior and trigger of gravity erosion, stability of different slope geometries under rainfalls, temporal changes of sediment concentrations and particle sizes before, during and after the occurrence of slope failures, and so on. This relentless and dedicated effort significantly advances our understanding of gravity erosion processes and mechanisms. The experimental results also provide a scientific basis for the optimization of gravity erosion control measures. However, I would encourage the authors to validate the experimental results using more field data collected under natural conditions in the future.

The check-dam is one of the most effective measures in soil erosion control on the Loess Plateau. Rainfall and runoff experiments have been performed to evaluate the check-dams' effectiveness in sediment detention and their role in stabilizing gully development and alleviating gravity erosion. The book also analyzes the long-term data collected from an experimental catchment to quantify the effectiveness of different biological and engineering measures in retaining water, soil, and nutrients, which provides practical guidance in the optimization of different conservation measures in the region. A decades-long comprehensive study on another much understudied but important erosion process in the Loess Plateau region, tunnel erosion, is presented in the book. This includes hydrological and sediment processes of tunnel erosion, initiation and development of tunnel systems, sediment contribution of tunnel erosion in the catchment, as well as geophysical methods in mapping tunnel systems in the field. The research has been internationally recognized.

Xiangzhou Xu, the first author of the book, received his doctoral degree at Tsinghua University ten years ago. During his graduate study, I found that Xiangzhou was a very diligent, self-motivated, and thoughtful student. What deeply impressed me is that he always earnestly accepted my advices and creatively enriched the ideas I told him. He conducted a series of soil conservation experiments and completed his doctoral thesis with distinction. I am pleased that Xiangzhou continues to show a great passion, perseverance, and unwavering determination to pursue his research interests after his graduation. Presently, he has deepened his researches and obtained many new academic achievements highly recognized by fellow academics in the soil conservation areas.

I applaud the great accomplishments the authors have made and I highly recommend this book to anyone who is in soil erosion or relevant field.

Academician Guangqian Wang is a full professor and doctoral supervisor at the Department of Hydraulic Engineering, Tsinghua University. He is granted with the National Natural Science Foundation of China for Distinguished Young Scholars in 1995, hired as a distinguished professor in Cheung Kong Scholars in 2000, and elected as an academician of the Chinese Academy of Sciences in 2009. Academician Guangqian Wang is also the president of Qinghai University, a standing committee member of the 13th Chinese People's Political Consultative Conference, and a vice chairman of the 12th Central Committee of the China Democratic League.

Academician Guangqian Wang is a leading scientist in the field of hydroscience and river governance. He presents a dynamic two-phase flow model of sediment transport based on the basic theory for flow and sediment transport, which reveals the different characteristics of water turbulence and particle movement. Moreover, he proposes a dynamic model of soil erosion and sediment transport in a basin, which can realize the process coupling of the river channel and basin, and extend the sediment research from the river to basin scale. Findings derived from these researches have advanced the field of sediment science. In addition, he has solved many key scientific and technological problems in engineering applications. For example, the key problems concerning the break-flow, river burst, and soil erosion in the Yellow River Basin, the problem of sediment in the Three Gorges, and the danger release of the barrier lake caused by the earthquake in the Yangtze River. Academician Wang has published numerous scientific monographs and won many awards including the first and second prize of the National Award for Progress in Science and Technology.

Guangqian Wang
Academician and Professor
Tsinghua University, Beijing, China

Foreword by John Zhang

This book is the first to systematically explore the experimental erosion by integrating theory, erosion observation, and conservation application. Although numerous books have been published on soil erosion both in English and in Chinese, none has concentrated on the experimental studies on the Loess Plateau of China, with an attempt of establishing a new sub-discipline: experimental erosion. The work included in the book represents exemplary studies in the field of soil erosion and conservation. The new methods and results of the research will also provide practical guidance for controlling soil erosion. Hence, the book will be valuable to both soil erosion researchers and conservationists.

The book is clearly written and well structured. First, the authors establish the theoretical framework and fundamental principles of experimental erosion. Second, they present a variety of state-of-the-art experimental techniques and innovative observation methods for designing soil erosion experiments. Finally, they demonstrate experimental case studies on a range of erosion problems and conservation practices such as mass movement, gully erosion, tunnel erosion as well as check-dam.

The work described in the book tackles several important, but difficult and understudied problems in soil erosion research. Rainfall simulation experiments have long been applied to study soil erosion processes, but largely limited to rain splash, sheet wash, and rill erosion mostly on gentle to medium slopes. In this book, several different types of rainfall simulators and a mobile laboratory, designed by Prof. Xu, were employed to study mass movements on steep slopes both in the laboratory and fields. A Topography Meter, also invented by Prof. Xu, was used to quantitatively observe the processes of mass failures during the experiments. The valuable data acquired by the rainfall simulation experiments are of critical importance in studying the mechanism and sediment production of mass movement, due to the difficulties in predicting its timing, location, and magnitude under natural conditions. Also, rainfall simulation experiments were conducted in the field to study gully formation and development, which have rarely been done elsewhere. Check-dams are among the most widespread engineering structures in conserving soil and water on the Loess Plateau. The efficiency and effectiveness of a check-dam system in retaining sediment and stabilizing gully development were investigated by a series of experiments using semi-scaled models as described in the book. Such research results are of tremendously practical significance in controlling soil erosion on the Loess Plateau of China.

Soil piping and tunnel erosion have not been widely studied as compared to surface erosion. Tunnel erosion on the Loess Plateau is the most serious ones in the non-karst areas in the world. Nevertheless, the hydrological and sediment processes have never been studied in-depth in the region. The work conducted in the experimental catchment, Yangdaogou Catchment, was the first attempt to monitor hydrological and sediment processes of tunnel flows on the Loess Plateau. The long-term continuous investigations of tunnel formation and its relation to gully development, mass movements, as well as the sediment and hydrological processes of tunnel flows represent the most comprehensive study of soil piping/tunnel erosion in semiarid areas in the world.

The authors have long worked in the soil erosion and the related fields and are outstanding scholars who have earned national and international recognitions. Dr. Xiangzhou Xu is a professor and Ph.D. supervisor at the Dalian University of Technology. He has published more than fifty articles and two books, as well as obtained thirteen patents for developing erosion monitoring and measurement devices or techniques. Dr. Tongxin Zhu is a professor at the University of Minnesota Duluth in the USA and has published over thirty articles in books and leading international hydrological and geomorphological journals. Dr. Hongwu Zhang is a distinguished professor at Tsinghua University and a leading scientist in the areas of river management and hydraulics. He has published nine books and over 200 journal articles and has received numerous prestigious awards and honors.

In conclusion, this book is a valuable piece of scholarship and will be a useful reference to the graduate students, soil erosion scientists and engineers, and soil and water conservationists. I am looking forward to its publication with great interest.

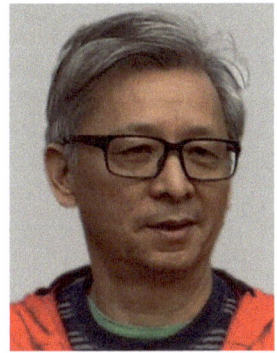

Xunchang (John) Zhang, a research hydrologist in the USDA-ARS, is nationally and internationally recognized for expertise in the field of erosion processes. Dr. Zhang developed and tested a new technique to generate spatially distributed erosion data using Rare Earth Elements (REE) as tracers. Since the publication of his work, the technique has been used by scientists from many countries and across many disciplines. Moreover, a downscaling method presented by him also overcomes the stationary limitation suffered in all existing approaches. Dr. Zhang has authored or co-authored over 100 SCI journal articles, one book, and seven chapters during his over 30 years of research. He received six best paper awards in four research subject areas. Several of his papers have had major impacts on advancing sciences in the field.

John Zhang

Dr. John Zhang
Research Hydrologist
Grazinglands Research Laboratory
USDA-ARS, El Reno, USA

Preface

Presently, soil erosion has become one of the most common environmental problems in the world. Experimenting is an important means to study the regularity and control measures of soil erosion. This book presents a new subject, EXPERIMENTAL EROSION, as a member of the discipline of soil and water conservation. Experimental erosion is defined to explore the transportation law and control method of soil and water loss by performing field or laboratory tests under closely monitored or controlled experimental conditions. The similarity theory, simulation and observation technology, and data processing method are the three pillars of the experimental erosion.

The book includes a chapter Introduction and other two parts. Chapter 1 is an overall introduction of the book mainly about the background and significance of this study together with the framework of the experimental erosion. Part I elaborates on the theories and methods for the soil conservation experiments, and Part II presents some representative case studies of soil conservation experiments. In Part I, Chapter 2 reviews the development of experiments on soil conservation at first and then puts forward the definition and scope of soil conservation experiments including the scale model, semi-scale model, segment of natural watershed, analog model, and response model. Then, an emphasis in the chapter is paid on the experimental method of a semi-scale physical model that can predict the amount of soil loss in a catchment. An applicable and controllable experimental device is crucial to obtain credible data in a soil and water conservation experiment. Chapter 3 contains a review of the simulation devices in soil conservation experiments and rainfall simulators designed by the first author and his students. Chapter 4 presents the measurement devices and data processing methods to observe the gravity erosions. Investigations of gravity erosion usually require special techniques and methods due to its unique origin and nature. The Topography Meter invented by the authors can be applied to quantitatively observe the process of mass failures on the steep slope. Chapter 5 demonstrates a portable landslide laboratory applicable for the remote regions and a few sets of rainfall simulators, which have gained over 10 patent authorizations.

In Part II, representative case studies on erosion research are exhibited, with the purpose of providing a scientific context of soil conservation in the semiarid regions, particularly on the Loess Plateau of China, as well as testifying the effectiveness and practicability of the above devices. Chapter 6 describes the statistical analysis and field investigation of landslides on the Loess Plateau. Data collected here may provide the prototype proofs to study soil conservation with field experiments. Chapter 7 calculates the impact factors of conservation practices on soil, water, and nutrients in a representative catchment of the Loess Plateau based on field experiments. The results show that soil erosion and nutrient loss had been greatly mitigated with various conservation practices, and check-dam was the most appropriate conservation practice on the Loess Plateau. In Chap. 8, a runoff simulation experiment for a single check-dam and a rainfall simulation experiment for the whole check-dam system were conducted. The experiments demonstrate a characteristic phenomenon, i.e., relative stability of the check-dam system on the Loess Plateau and evaluated the effects of the dam construction sequence on the soil deposition in the catchment. The gravity erosion is a dominant geomorphic process on the steep loess slopes. In Chap. 9, rainfall simulation experiments were conducted to monitor the occurrence and behavior of the mass failure on steep loess slopes. The results show that the avalanche played a crucial role of soil erosion as the landform was made with loess by hand patting. Chapter 10 describes a series of laboratory experiments aiming to test the stability of different slope geometries under rainfalls. A sensitivity analysis is also performed to quantitatively explore the triggering mechanisms of mass failures on the steep loess slope. In Chap. 11, a series of experiments in the laboratory were conducted to observe the slope geometries after mass failures, and then a sensitivity analysis was performed to quantitatively explore the triggering mechanisms of failure scars on the steep loess slope. Chapter 12 introduces the rainfall simulation experiments conducted on the natural loess slopes where a series of mass failures happened under the simulated rainfall. An index of enrichment/dilution ratio was used to quantitatively explore the change of suspended sediment sizes affected by gravity erosion. In Chap. 13, detailed monitoring of water and sediment delivery from the major tunnel systems was conducted in a small semiarid sub-basin, locally known as the Yangdaogou Catchment, in the hilly loess region of the Loess Plateau.

Beginning in 2001, a series of experimental studies were performed by the first author together with his students and partners to generate and test the hypotheses for the mechanism of the soil erosion on slopes and gullies, sediment storage effects of check-dam systems, and behaviors and triggers of gravity erosion on the steep slope. Some papers have presented the results of these experiments, but much of the work remains unpublished. The characteristic features that make the study distinctive are shown as follows: (1) The theory system, i.e., the three pillars of the Experimental Erosion, is presented in this book. The theoretical concepts presented here are those needed for an understanding of soil behavior in the laboratory/field tests described in subsequent chapters. A semi-scale model for simulating soils retained by check-dams, the increase-rate-analysis method to evaluate the sensitivity of the erosion to the causal factors, a moveable laboratory for field tests, a

Topography Meter to measure gravity erosion on gully bank, and several new type of rainfall simulators suitable for laboratory or field experiments have been presented in this book. (2) Behavior and trigger of gravity erosion on the steep loess slope are the focus of the authors' recent research. As we know, the experimental work on the sheet erosion or rill erosion has been extensive, but much less has been accomplished regarding gravity erosion. (3) The relative stability theory and optimization method for the check-dam system have been verified via the semi-scale model experiment. In practice, the check-dam system is the most effective way to control gravity erosion on the gully bank in the catchment of the Loess Plateau.

One main objective of this book is to show the monitoring and modeling method for soil scientists who design and prosecute experimental studies of soil loss. The new data collected in the tests aforementioned can be beneficial to protection and remediation of soils. Thus, another objective of this book, which is the most important, is to make the results of these experiments more generally available. We have attempted to assemble and integrate the experimental results, both published and unpublished. In-depth discussions of the experimental data with the new methods of data processing, e.g., the increase-rate-analysis method of analyzing the sensitivity of gravity erosion, are also presented and used. Here lies the basic reason why this new book is called "Experimental Erosion," in which not only experimental aspects are considered but also theoretical and analytical aspects in relation to the soil conservation. The book will be envisaged as an overview of the future research mentioned earlier.

The published book is sponsored by Dalian Municipal Government. Prof. Xiangzhou Xu conceived, designed, and drafted this book. Prof. Tongxin Zhu wrote Chap. 13 and co-authored Chap. 12. Professor Hongwu Zhang participated in the writing of Chaps. 2 and 3. Doctoral student Lu Gao participated in the editing and proofreading of the whole manuscript. We sincerely thank Prof. Guangqian Wang for his continuous encouragement and support in the research works related to the book. We want to thank the Professors, Tingwu Lei, Chongfa Cai, Yongming Shen, Junqiang Xia, and Yongqiang Cao, who have given valuable advices for revising the book. In addition to former students and colleagues who permitted us to use the product of their research efforts, the authors acknowledge the support of the programs for National Key R&D Project (2016YFC0402504), entitled Management of the River Bed and Bench Land in the Lower Reaches of the Yellow River; General Programs of the National Natural Science Foundation of China (51879032; 51179021; 51079016), entitled Landslides on the soil reservoir bank for the different factors coupling processes, A Field Study on Gravitational Erosion of the Loess Slope and An Experimental Study on Gravitational Erosion of the Loess Slope, respectively; Excellent Talents in the Universities of Liaoning Province (LR2015015), entitled Experimental Erosion; and Open Research Fund of the State Key Laboratory of Soil Erosion and Dryland Farming on the Loess Plateau (A314021402-1609), entitled Triggering, Processes, and Sensitivity Analysis of the Gravity Erosion on the Loess Gully Bank. Chapters relating to the research work performed by the graduate students have been reviewed by Wenzhao Guo, Yulei Ma, Xingyang Zhao, Junwen Yin, Jiyuan Lu, Zhenyi Liu, Qiao Yan,

Chao Zhao, Feilong Xu, Lu Yu, and Jiangli Guo. We would like to thank all the people from Science Press and Springer Nature who were involved in the publication of this book.

There is an interesting saying that no one believes a hypothesis except its originator but everyone believes an experiment except the experimenter, because to transfer the experimental results through the analogy to larger landforms is so difficult. Confirmation of the experimental results must come from field investigations. Hence, more field experiments are anticipated with the precise observation of the process and strict control of the boundary in the near future. Another limitation of the book is that most of the authors are scientists from China. I am afraid that language problems still exist although we have done a great endeavor. We sincerely welcome any comments and criticisms from those who use it.

Dalian, China   Xiangzhou Xu
April 2020

Contents

About the Authors

Prof. Dr. Xiangzhou Xu is a full professor and Ph.D. supervisor at Dalian University of Technology (China). His expertise includes resource sustainability, soil and water conservation, urban rainwater utilization, and topography measurement. His research focus lies in physical experiments in the laboratory as well as in the field. The Topography Meter developed by him opened a new way for further tracking and measuring the landslides. He is the principal investigator of over ten national, ministerial and provincial projects, including three projects funded by the National Natural Science Foundation of China and one by the Excellent Talent Programs in the Universities of Liaoning Province. He has also won two prizes of the Ministerial and Provincial-Level Science and Technology Awards as the first or third accomplisher. He is the first or corresponding author of over 50 journal articles, and first or co-author of three monographs. He has filed 2 PCT patents, and more than 20 Chinese invention patents (16 of which have been granted).

Prof. Dr. Tongxin Zhu is a full professor at the University of Minnesota Duluth (USA). His research interests focus on the hydrological and geomorphic processes at the small watershed scale by integrating field investigations into simulation modeling. He is internationally recognized for his prominent studies on tunnel erosion in semiarid climatic conditions. He has published 20 SCI-indexed papers in a range of international journals.

Prof. Dr. Hongwu Zhang is a distinguished professor and Ph.D. supervisor at Tsinghua University (China). He is also a leading scientist in the areas of river management and hydraulics. Presently, he has authored nine books and over 200 journal articles. He is the principal investigator of a National Key R&D and a Key Project of the National Natural Science Foundations of China, and he also won two second prizes of National Scientific and Technological Progress Award as the third author.

Ms. Lu Gao is a Ph.D. student supervised by Professor Xiangzhou Xu at Dalian University of Technology, China.

Chapter 1
Introduction

Abstract An understanding of erosion processes is important for the design and operation of soil conservation projects, and study related to environmental and ecological issues. This collection of work develops new data sets and experimental methods to quantify the dynamics of soil loss that represent different stages in the development of soil functions. The application of the new data tests how simulation and observation can be coupled to guide beneficial intervention in soils in order to control soil erosion, especially that on the steep slope.

Keywords Background · Similarity law · Rainfall simulation · Observation technique

The study of soil erosion may focus on experimental science for the controlled experiment lays a firm foundation for soil-conservation research (Zhao et al. 2012; Guo et al. 2017). In order to understand a large, complex, and slowly evolving geomorphic system, earth scientists may resort to the study of a smaller and simpler analog (Schumm et al. 1987). Chorley (1964) identified three broad classes of physical models, namely, segments of unscaled reality, scale models, and analog models. The scale model may be a replica of a natural landform, scaled in such a way that ratios of significant dimensions and forces are equal to those in nature, although their absolute magnitudes may be greatly different. Scale models need not necessarily to reproduce a specific prototype, but they may be a scaled-down version of a general class of geomorphic features. Furthermore, a physical analog model may reproduce some significant aspects of the form and function of a natural phenomenon, but the forces, materials, and processes may be quite dissimilar to those in nature. Physical models can be used in generating and testing for hypothetical, predictive, and descriptive purposes. However, the inherent uncertainties in the relationships between the model and their prototypes always require field validation before the results can be utilized with confidence for prediction or postdiction. The monitoring of segments of unscaled reality is a widely used approach, and it has a long history. A particularly fruitful use has been adopted by the agricultural engineers who collect runoff and sediment yield data from field plots with different soils, cover types, and management regimes.

A variety of techniques have been used to estimate the soil erosion in previous studies, each with intrinsic limitations and uncertainties. Statistics on the flow characteristics in the process of confluence in various gullies revealed that the combination of sediment concentration and size distribution played a critical role in maintaining the flow with heavy sediment load owing to the natural adjustments in the Loess Hill Ravine Region of the Yellow River Basin (Wang et al. 1982). With a similar method, Hovius et al. (1997) also found that sediments discharged from the western Southern Alps were dominated by landslide-derived materials. The strain probe method was used to continuously detect soil creep *in situ*, and then the volume of slope failure was accordingly pre-estimated (Iverson et al. 2000; Yamada 1999). The tracer element method is a helpful approach (Wen et al. 2003). Stereo photogrammetry was also applied to determine the surface movement of the failed landslide mass (Ochiai et al. 2004). Recently, remote sensing technologies, including terrestrial laser scanning (Oppikofer et al. 2008), sonar bathymetry (Haflidason et al. 2005), radar altimetry (Velicogna and Wahr 2006), aerial photography (Martin et al. 2002; Whitehouse 1983), approach combing aerial photograph with satellite imagery (Li et al. 2013), have been used to monitor soil erosion and geomorphic evolution. However, it is very difficult to monitor the time-varying process of an individual mass failure with these techniques because of the randomness and suddenness of such an event.

Rainfall simulation is widely used for the experimental study on soil and water conservation. The simulators could be divided into two main groups (Cheng et al. 2008): the non-pressurized rainfall simulators, including the "thread droppers" and the "needle droppers" with drop formers primarily made of wool fibers or hypodermic needles, and the pressurized water rainfall simulators, including the "spout" and the "sprayer", of which raindrops sprayed out of a row of spouts on the pipe or a nozzle under pressure. The former could be hardly found at present due to its drawback that the minimum size of drops produced is far larger and more even than most of the natural rainfalls and the simulator could not easily generate rains with kinetic energy similar to the natural ones. Rainfall simulation halls with automatic operation and observation systems have been booming in recent years, but the basic simulators will not be completely replaced in virtue of the cheap price and convenient manipulation.

Nevertheless, the main limitation of the experimental method is probably linked with the major difficulty to transfer the experimental results by analogy to larger landforms (Schumm et al. 1987). Confirmation of the experimental results must come from field investigations, and frequently this can be done by considering existing field data from a new perspective that has been provided by the observation of the small landforms as they evolve and react. Hence, what we could do is to simulate the landslide processes by employing a conceptual slope under rainfall simulation. However, the representative catchments on the Chinese Loess Plateau for scientific research are generally far from urban areas, lack water and electricity, and have deep ditches, steep slopes, inconvenient transportation, and strong winds. The special situation poses another severe challenge to the site survey. Here a mobile shelter is recommended in the book that could be assembled in the field to conduct site-specific tests based on the terrain of the Loess Plateau.

This collection of work includes a wide range of types of research and applied studies undertaken using this experimental approach and, as such, serves to demonstrate the value and effectiveness of adopting such an investigative means. In spite of the significant problems associated with the design and prosecution of experimental studies of soil conservation, they can provide an insight into landform evolution and dynamics that can be obtained in no other way. The work develops new data sets and experimental methods to quantify the dynamics of soil loss that represent different stages in the development of soil functions. In conclusion, application of the new data tests how simulation and observation can be coupled to guide beneficial intervention in soils in order to control soil erosion, especially that on the steep slope.

References

Cheng F, Xu X Z, Gao J H, et al. 2008. Advance of research on the rainfall simulators for soil erosion. Science of Soil and Water Conservation, 6(2): 107–112 (in Chinese).

Chorley R J. 1964. Geography and analogue theory. Annals of the Association of American Geographers, 54(1), 127–137.

Guo Z, Ma M, Cai C F, et al. 2017. Oil erosion and flow hydraulics on red soil slope under simulated rainfall/runoff, Resources and Environment in the Yangtze Basin, 26(1): 150–157 (in Chinese).

Haflidason H, Lien R, Sejrup H P, et al. 2005. The dating and morphometry of the Storegga Slide. Marine and Petroleum Geology, 22(1–2): 123–136.

Hovius N, Stark C P, Allen P A. 1997. Sediment flux from a mountain belt derived by landslide mapping. Geology, 25(3): 231–234.

Iverson R M, Reid M E, Iverson N R, et al. 2000. Acute sensitivity of landslide rates to initial soil porosity. Science, 290(5491): 513–516.

Li Y G, Chen G Q, Wang B, et al. 2013. A new approach of combining aerial photography with satellite imagery for landslide detection. Natural Hazards, 66(2): 649–669.

Martin Y, Rood K, Schwab J W, et al. 2002. Sediment transfer by shallow landsliding in the Queen Charlotte Islands, British Columbia. Canadian Journal of Earth Sciences, 39(2): 189–205.

Ochiai H, Okada Y, Furuya G, et al. 2004. A fluidized landslide on a natural slope by artificial rainfall. Landslides, 1(3): 211–219.

Oppikofer T, Jaboyedoff M, Keusen H R. 2008. Collapse at the eastern Eiger flank in the Swiss Alps. Nature Geoscience, 1(8): 531–535.

Schumm S A, Mosley M P, Weaver W E. 1987. Experimental fluvial geomorphology. New York: John Wiley and Sons: 1–7.

Velicogna I, Wahr J. 2006. Measurements of time-variable gravity show mass loss in Antarctica. Science, 311(5768): 1754–1756.

Wang X K, Qian N, Hu W D. 1982. The formation and process of confluence of the flow with hyperconcentration in the Gullied-Hilly Loess Areas of the Yellow River Basin. Journal of Hydraulic Engineering, (7): 26–35 (in Chinese).

Wen A B, Zhang X B, Zhang Y Y, et al. 2003. Comparison study on sediment sources between debris flow gullies and non-debris flow gullies by using the 137Cs tracing technique in Dongchuan, Yunnan Province of China. Journal of Sediment Research, (4): 52–56 (in Chinese).

Whitehouse I E. 1983. Distribution of large rock avalanche deposits in the central Southern Alps, New Zealand. New Zealand Journal of Geology and Geophysics, 26(3): 271–279.

Yamada S. 1999. The role of soil creep and slope failure in the landscape evolution of a head water basin: field measurements in a zero order basin of northern Japan. Geomorphology, 28(3–4): 329–344.

Zhao C, Cai C, Ding S, et al. 2012. Design of Minitype Flume for Simulation Experiment on Soil Erosion. Agricultural Engineering, 2(1): 64–66 (in Chinese).

Part I
Theories and Methods for Soil Conservation Experiments

A successful experimental model with correct theories and appropriate methods can provide an insight into landform evolution and dynamics.

Chapter 2
Similarity of Model Experiments

Abstract A successful experiment with a physical model requires necessary conditions of similarity. This chapter presents an experimental method with a semi-scale physical model. Four criteria are mentioned here: similarities of watershed geometry, grain size and bare land, Froude number (*Fr*) for rainfall event, and soil erosion in the downscaled models. The efficacy of the proposed method was confirmed using these criteria in two different downscaled model experiments. Experimental results show that while the amount of soil loss in the small scale models was converted by multiplying the scale number, the amount was very close to that of the large scale model. Obviously, with a semi-scale physical model, the experiments are available to verify and predict the soil loss in a catchment with the check dam system on the Loess Plateau, China.

Keywords Loess Plateau · Check dam · Semi-scale model · Similarity condition · Soil loss

2.1 Development of the Experimental Erosion

Different from the monitoring or investigating method, an experiment of soil conservation under closely monitored or controlled experimental conditions focuses on the regularities of soil loss from a micro perspective. According to the formation of the underlying surface, the physical models may be categorized into two broad classes, namely, *in situ* tests and laboratory experiments. Runoff and sediment distribution in a segment of the natural watershed may be conducted in an *in situ* test, i.e. field experiment, under natural and simulated rainfall conditions to study the mechanism of soil and water loss and the effects of management measures. As a kind of representative field experiment, the runoff plot has been widely used all over the world. The landform is not reconstructed although the border of the experimental area should be strictly restricted in a field experiment. The processes of soil and water loss under a complicated topographical condition may be relatively easily simulated and observed in the field, because no change exists in the scales of the underlying surface and properties of the erosion material. In contrast, a laboratory experiment, i.e. model experiment, is performed in a scaled underlying surface reconstructed with

© Science Press and Springer Nature Singapore Pte Ltd. 2020
X. Xu et al., *Experimental Erosion*,
https://doi.org/10.1007/978-981-15-3801-8_2

the specific materials. Both of the natural or simulated rainfall, mostly the simulated one, may be applied in a model experiment. The laboratory experiments include full-scale models and variational-scale models. In a full-scale model experiment, the geometry scale of the underlying surface is consistent with the prototype, while in a variational-scale model, the landform is scaled down based on the prototype. A successful physical model experiment requires similar necessary conditions. However, the model experiment may be conducted in the laboratory far away from the study area, and the experimental scenario may be freely designed if needed. Hence, the model experiment is recommended as an important supplement of the field experiment and some advantages of the experiment in the laboratory cannot be substituted with field experiments.

These can be categorized into field experiments and model experiments according to the formation of the underlying surface of the experimental area. Based on characteristics of the watershed geomorphology, runoff and sediment distribution under natural and simulated rainfall conditions or other dynamic conditions of erosion, field experiments are conducted to study the mechanism of soil and water loss and the effects of management measures on soil and water conservation, such as the small-plot runoff experiments. It is easy to simulate and observe the complicated topographical conditions of soil and water loss in the field, because there is no change in the scale of the underlying surface and properties of erosion material, which means that to reconstruct the plot is not needed. Based on geomorphology and geology of the natural watershed, model experiments are conducted to study the mechanism of soil and water loss and effects of various management measures for the prototype watershed under the natural or simulated rainfall events (mostly simulated rainfall events) by reconstructing the plot using certain materials. Model experiments include the full-scale model experiments and the scale model experiments. For a full-scale model experiment, the geometry scale of underlying surface is consistent with its prototype, while the scale model experiment is designed based on the zoom scale of prototype. A successful physical model experiment requires similar necessary conditions. However, model experiments can be conducted in the laboratory far away from the soil loss field and assumed with an unlimited scale of the underlying surface. Therefore, some features of the model experiments cannot be replaced by those of the field experiments, and the model experiments are recommended as an important supplement to the field experiments.

Under laboratory conditions, measurements are more accurate and many experiments can be conducted (Cerdà and García-Fayos 2002). Thus, downscaled models are currently common in many different engineering fields, such as hydraulics and river engineering. The advantages of such models are well known (Zhang 1994). However, few studies have simulated the process of soil loss using downscaled model experiments, because simulating the hydrological, morphological, and geological conditions is extremely complex. Nevertheless, dimensional analysis links various observed phenomena of erosion and deposition into a unified process, and enables complete predictions in the landfrom changes to be anticipated when watershed treatment is altered (Strabler 1958). Downscaled models are typically used to simulate physiognomy performance (Jin et al. 2003; Hancock and Willgoose 2003). Hancock and Willgoose (2004) investigated the effect of erosion on a back-filled and a capped earthen dam wall by constructing an experimental landscape simulator in the laboratory. Due to the design of the rainfall simulator, it is difficult to directly scale the rainfall-runoff processes to the field. Consequently, no attempt has been made to match the rate of gully development on the tailings dam to field-scale processes.

Recently, downscaled model experiments on soil erosion in small watersheds of the Loess Plateau have been performed, and good progress has been made in similarity methodology. Shi et al. (1997a, b) observed the quantitative erosion in the gullies and on slopes of the downscaled watershed model of the Xiaofanjiagou Gully, Shaanxi Province. However, the proportions of soil erosion in gullies and on slopes in the model were not very reliable, because the model runoff was underdeveloped and erosion types were changed, compared with those in the prototype due to insufficient similarity in rainfall dynamics and ground cover. Jiang et al. (1994) and Yuan et al. (2000a, b) performed a series of downscaled model experiments for different degrees of erosion control in the small watersheds on the Loess Plateau to evaluate the relationship between the runoff and soil loss, in which the explicit similarity conditions to those in river engineering were considered. These experiments were relatively successful; however, some aspects of the model design are still in dispute (Zhang and Zhang 2000). The Chinese government has paid considerable attention to the theory underlying scale models for soil conservation on the Loess Plateau. The theory on model-based Loess Plateau has been proposed, which includes prototype, digital model and physical model (Li 2001). Several national foundations have been given to develop the theory designing the physical models, especially the scale models, which could simulate the soil erosion processes in the small watersheds on the Loess Plateau.

The Loess Plateau is characteristic for the large area, complex geomorphology, and considerable soil erosion. Many gullies are required to reduce soil and water loss. Constructing check dams in the gullies is an effective strategy for reducing sediment loss. More than 100,000 check dams have been built over the last 50 years on the Loess Plateau; however, few analogous dams have been constructed in other countries. The Chinese Ministry of Water Resources states that 163,300 check dams will be built on the Loess Plateau by 2020 (MWRC 2003). Designing check dam systems requires the estimates for (1) preferred dam sites, (2) the number of dams required and their heights for sediment retention and flood control, and (3) the optimal sequence and interval for dam construction. In an optimum design for the dam-construction, the maximum amount of sediment can be retained by check dams, whereas the following features are determinate: the number, location and capacity of check dams for a catchment, and the erosion rate for the controlled area (Tian et al. 2003). However, check dams have not been sufficiently discussed in the international literature. This study presents a novel experimental method for assessing the soil retained by check dams in the small watersheds on the Loess Plateau, China.

2.2 Purpose and Significance

The proposed method predicts the sediment retained by check dams based on rainfall, land cover and geological conditions in the prototype catchment before the check dams are constructed. Suppose that a small watershed on the Loess Plateau would be managed, and rainfall-erosion data before construction of the check dams were available. To predict the erosion while a check dam system is being constructed, we assume that other erosion factors, e.g., rainfall, erosion material, and plant cover, would be similar to those before the check dams were constructed.

The researchers should consider the similarity of the cumulative effect while designing the downscaled model experiment for soil conservation. For the erosion controlled by check dams, the cumulative effect for at least one year by the check dam may be considered rather than the motion process of each soil particle. In fact, the annual sediment yield in the small watersheds on the Loess Plateau is primarily generated by a few strong rainstorms; that is, the short and intense rainfall events are responsible for 60–90% of the total soil loss over a 1-year period. Since the cumulative effect of check dams to retain sediment is of concern and not the temporal evolution of the slopes or gullies, a downscaled model experiment must ensure that the ratio of the prototype soil loss after rainfall events to model soil loss after the same rainfalls, namely the scale number of the soil loss, remains constant. In this instance the scale number of the soil loss could be pre-calibrated. Thus, the amount of soil loss in the model watershed after constructing check dams represent the prototype soil loss. In this study, four criteria are utilized to ensure that the scale number of the soil loss remains constant. Firstly, the initial dimensions of the landform required by the geometrical similarity, including those of the check dams, are scaled down according to the dimensions of the prototype watershed with the same proportion in the horizontal and vertical orientations. Secondly, for the similarity of erosion form, the simulated soil similar to that of the prototype is used, and the model rainfall intensity exceeds the soil erosion threshold. Thirdly, the rainfall duration was determined by the Froude number (Fr) similarity. Finally, for the similarity of the rainfall-erosion, the relationship between the rainfall and erosion in the model experiment corresponds to that of the prototype. The proposed method provides the quantitative proportion of soil loss between the prototype and model using similarity criteria. Nevertheless, it does not strictly meet the similar situation in the conventional scale model experiment. Thus, the proposed method is defined as the semi-scale physical model experimental method (SSPM).

The SSPM was tested using two downscaled experiments (Fig. 2.1). A large scale model, the Model B, at 1:60 of the prototype landform, was used as the simulated prototype watershed. Two small scale models, Model Da and Model Db, with different erosion rates but the same geometrical size, each at 1:240 of the prototype landform, were employed to simulate the hydraulic process in Model B. Data from these three models are comparable because the experimental devices and observation measures in the small scale models are consistent with those in the Model B. Moreover, an

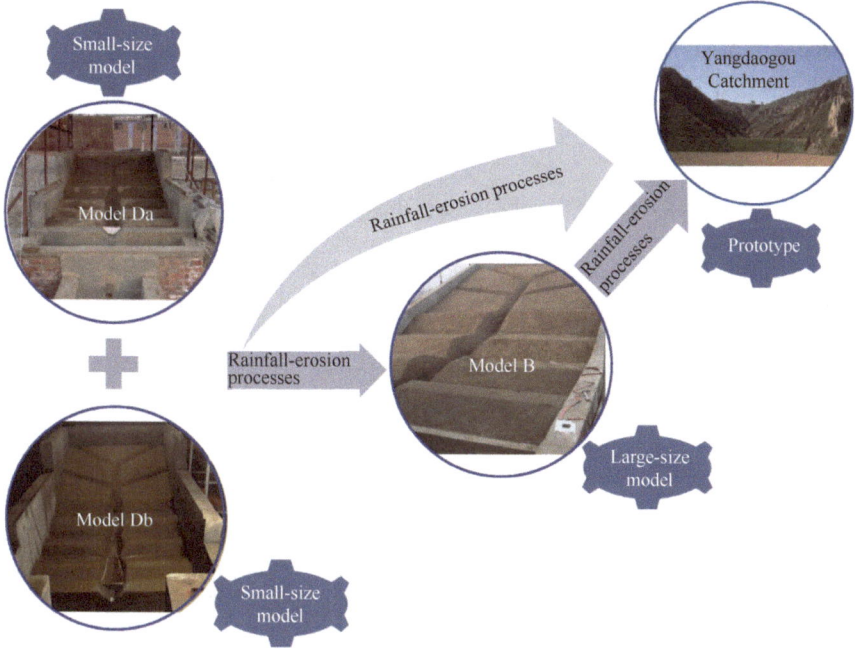

Fig. 2.1   A technical framework to verify the SSPM

across-the-board comparison was possible as the soil erosion processes in the three models were all implemented in the laboratory using simulated rainfall.

2.3   Similarity Requirements

The similarity degree between the prototype and model is based on the geometrical, kinematic and dynamic similarity. Geometrical similarity in a model may be achieved when each geometrical length L_p in the prototype has a constant ratio corresponding to the model length L_m in the model. Subscripts p and m stand for prototype and model parameters respectively. This ratio is called the length scale number λ_L of the model ($\lambda_L = L_p/L_m$) In the proposed downscaled model experiment, horizontal length scale is equal to the vertical length scale; therefore, the dam height scale number λ_H equals the length scale number. Likewise, due to geometrical similarity, the watershed area scale number λ_A could be calculated as λ_L^2 where the subscript A is the area of the watershed.

Dynamic similarity indicates that the corresponding forces in the prototype and model must have a constant ratio. For the flow under its own gravity, e.g., the free surface flow, requires geometrical similarity and equality of the Froude number (Fr) in the model and prototype; namely $\lambda_{Fr} = 1.0$. In a field observation or laboratory experiment, the gravity effect is greater than viscosity for turbulent flows caused by the rainfall. Consequently, the time scale number of the rainfall duration, λ_t, is related to Fr similarity; that is,

$$\lambda_t = t_p/t_m = \lambda_L^{0.5} \tag{2.1}$$

where t is rainfall duration (s); subscripts p and m stand for prototype parameter and model parameter respectively.

The emphasis of the proposed method is that the ratio of geomorphologic evolvement ratio of the model corresponds to the prototype ratio, which remains constant after rainfall events, such that the prototype soil erosion processes can be measured based on the model experimental results:

$$\frac{\bar{Y}_{m(i)}/L_m}{\bar{Y}_{p(i)}/L_p} = \frac{\lambda_L}{\lambda_Y} = R \tag{2.2}$$

where i is the sequence of rainfall events; $\bar{Y}_{(i)}$ is the mean erosion/deposition depth of the landform during rainfall events (m); L is the length of the small watershed (m); Y/L is a dimensionless term, which stands for the geomorphologic evolvement ratio of a single rainfall event; and R is the ratio of erosion extent between the model and prototype. When $R = 1$, the erosion/deposition depth scale number, λ_Y, equals the length scale number, λ_L, illustrating that the erosion/deposition extent $\bar{Y}_p/L_p$ for a single simulated rainfall event in the prototype is equal to the product of the scaled model $\bar{Y}_m/L_m$. However, when $R \neq 1$, a model is distorted because the erosion/deposition scale differs from the length scale of the physical soil erosion model. Thus, the length scale number, λ_L, in the scaled model can be obtained by multiplying the erosion/deposition depth scale number, λ_Y, by ratio R.

Since the soil density ρ_s (kg/m^3) in the prototype is close to the model one, the soil loss scale number λ_S could be calculated as follows:

$$\lambda_S = \frac{S_p}{S_m} = \frac{A_p \times \bar{Y}_p \times \rho_{S_p}}{A_m \times \bar{Y}_m \times \rho_{S_m}} = \frac{\lambda_L^3}{R} \tag{2.3}$$

where S is the soil loss, and A is the drainage area. When $R = 1$, the scale number of soil loss, λ_S, equals the volume-scale number, λ_L^3, which could estimate the volume of the prototype soil loss based on the volume-scale number in the model experiment. When $R > 1$, the model erosion volume must be multiplied by the coefficient λ_L^3/R to estimate the amount of soil loss in the prototype, and the model landform varies more serious than the prototype landform under the same rainfall scale. When $R < 1$, the model erosion volume must be multiplied by a large coefficient to estimate the amount of soil erosion in the prototype. A comparison of the sediment transport capacity and sedimentation similarity in the river engineering indicates that the conditions of kinematic similarity and dynamic similarity are embodied in the scale of cumulative erosion volume for each rainfall event (Eq. 2.3). Four measures are required to ensure that the ratio R is constant for the semi-scale model: (1) The bare-land model with a corresponding erosion rate is employed in the experiment. (2) The dimensions of the landform, including the check dams, are scaled down according to the prototype watershed with the same proportions in the horizontal and vertical orientation. (3) The soil is similar to that in the prototype. (4) The initial water content in the soil layer before each simulated rainfall event remains constant.

In this study, if the ratio of the geomorphologic evolvement in the small scale model to the corresponding ratio in the large scale model remains constant after rainfall events, then R_D/R_B is a constant:

$$\frac{R_D}{R_B} = \frac{\bar{Y}_{D_{(i)}}/L_D}{\bar{Y}_{p_{(i)}}/L_p} \cdot \frac{\bar{Y}_{p_{(i)}}/L_p}{\bar{Y}_{B_{(i)}}/L_B} = \frac{\bar{Y}_{D_{(i)}}/L_D}{\bar{Y}_{B_{(i)}}/L_B} = R_{DB} \qquad (2.4)$$

where R_{DB} is defined as the relative ratio of erosion extent between the small scale model and the large scale model; in short, relative ratio. In this study, two small scale models with $R_{DB} \approx 1$ and $R_{DB} \neq 1$ were used to verify the SSPM, respectively.

2.4 Experimental Methods and Materials

The experiments were performed in the laboratory of the Yellow River Research Center, Tsinghua University, Beijing, China. Three experimental model landscapes, a large scale model (Model B) and two small scale models (Model Da and Model Db) were developed. (1) Each model was a small watershed that included the elemental geomorphologic units, e.g., gully, platform, and hill slope, with the representative characteristics of the small watersheds on the Loess Plateau. Each of the small scale models was a miniature version of the large scale model. (2) The loess soil was used as an erodible material in both models. (3) The same instruments were used for measuring the rainfall, soil loss and runoff in all models to ensure comparability of the data.

2.4.1 Data for the Prototype Watershed

Most areas on the Loess Plateau are arid or semiarid. Average annual precipitation is 350–550 mm, decreasing gradually from the southeast to the northwest. Most precipitation falls during the rainy season (June-September). A single rainstorm could account for 60–90% of the total annual precipitation, causing most soil loss in a given year. Rainfall events often last 30–120 min, resulting in runoffs with an average sediment concentration of 200–300 kg/m^3 and a maximum concentration of 1000 kg/m^3 (Wang and Jiao 1996). The Loess Hill Ravine Region is located in the middle and west of the Loess Plateau. Undulating terrain in the Loess Hill Ravine

Region is characterized by the crisscrossing gullies covered with the thick loess on top of ancient landforms. Ridges and mounds are the most typical micro-relief form on the Loess Plateau.

The prototype watershed, the Yangdaogou Catchment, is a typical small watershed located at the Loess Hill Ravine Region, and covers an area of 0.206 km^2 (Xu et al. 2006). The ground surface in the watershed is covered with Lishi loess and Malan loess (Cai et al. 1998). The major landform parameters, such as the drainage area, watershed elevation, hillside gradient, and length of the main gully, are all scaled down according to the dimensions of the Yangdaogou Catchment with the same horizontal and vertical proportions. Precipitation in the Yangdaogou Catchment is typical of that on the Loess Plateau. In the 1960s, when no dam existed, the average annual runoff rate was 36,700 m^3/(km^2 a), and the average annual soil erosion rate was 20,811 t/(km^2 a) (Zhang et al. 1995). Therefore, the annual soil loss, namely, the soil loss of the conceptual prototype in a rainfall event, S_p, could be obtained by multiplying the watershed area with the annual soil erosion rate, which is 4.29 × 10^6 kg/a. Generally, the annual erosion rate is adopted as an index of the structural practice while the check dam system is designed, and a rainfall event in the conceptualized prototype watershed accounts for the amount of soil loss in one year in the prototype watershed.

2.4.2 Landscape Simulator

The landscape simulator for the Model B consisted of a rainfall simulator suspended above a flume containing the large scale watershed model. The other two landscape simulators for the small scale models, the Model Da and Model Db, of which relative ratio R_{DB} was not equal to 1 and was close to 1, respectively, were used to verify the semi-scale model. The experimental apparatus and techniques in this study have been used extensively to examine the behavior of the watershed at a model scale, and could be used to generate the model watersheds with many features that are similar to field watersheds (Xu et al. 2006). The determinant parameters of the three models are listed in Table 2.1.

Table 2.1 shows the determinant parameters of the prototype watershed. The table also lists the erosion data for the prototype watershed together with those for the large scale model and the small scale model, and similarity conditions.

Figure 2.2 presents a schematic representation of the experimental setup for the Model B. Two lines of five SX2004 Sprayer-styled Rainfall Simulators were utilized to simulate the rainfall in the experimental plot, which measured 6.0 m × 10.8 m. The simulator contained a nozzle with an inner rotor, which was screwed onto the top of the pipes and sprayed downwards to the landform located 5.5 m below. Figure 2.3 presents a schematic representation of the experimental setup for the Model Db. A line of three SX2004 Sprayer-styled Rainfall Simulators was used to simulate the rainfall in the experimental plot, which measured 1.5 m × 2.7 m. Figure 2.4 presents a schematic representation of the experimental setup for the Model Da. A SX2002

Table 2.1 Experimental parameters

Variable	Prototype	Model B	Model Da or Model Db	Comments
Length of the main gully, L (m)	752	12.5	3.1	Restricted by the experimental ground
Dam height, H (m)	*	$\frac{1}{60} H_P$	$\frac{1}{240} H_P$	Geometrical similarity
Drainage area, A (m^2)	2.1×10^5	60.2	4.1	Geometrical similarity
Rainfall duration, t (min)	60–240	20	10	Froude number similarity
Soil concentration, C (kg/m^3)	200	50–70	50–70	$\lambda_C = 1.15$–3 (Zhang et al. 1994)
Soil loss, S (kg)	4.29×10^6	121.5	4.75	Distorted-scale by Eq. (2.3)
Particle density, ρ_s (10^3 kg/m^3)	2.65	2.56	2.56	–

*Dams 1, 2 and 8 were 18 m high, Dam 7 was 9 m high, and Dams 3, 4, 5, 6, 9, 10, 11, 12 were 5.4 m high

Spout-type Rainfall Simulator, comprising several rows of leptosomatic PVC pipes with the spouts 1 mm in diameter, were horizontally arranged on the top of a metal frame 4 m above the Model Db landscape. The simulator was utilized to simulate the rainfall in the experimental plot, which measured 1.5 m × 2.7 m.

Water was pumped from a nearby reservoir to a constant head tank, and then pumped to the rainfall simulators. A short but very intense downpour was applied, representing the typical rainfall events on the Loess Plateau. Sprinkling intensity was adjusted by pressure valves to maintain a constant hydraulic head for all nozzles located at different elevations. A flow-meter and a cut-off valve for each rainfall unit were manipulated manually to adjust the delivery rate of an immersed pump for the desired rainfall intensity. The rainfall distribution uniformity, determined using rain gauges spaced equidistantly over the landscape surface, was measured for three separate 10 min periods before the experiments started. The uniformity coefficients of the rainfall intensities for the three simulators exceeded 80% and were constant with time, especially when the rainfall intensity was 1.2–2.5 mm/min for the Model B and 1.3–3.5 mm/min for the Model Da and Model Db. Drops, with a median diameter of about 1.2–2.0 mm, were produced in the simulators and measured by catching the drops on a sheet of absorbent paper. The rainfall-erosion is associated with the energy, intensity and uniformity of the rainfall. The average energy impacting the ground cover per unit area and per unit time of the rainfall generated by the SX2004 Sprayer-styled Rainfall Simulator on the Model B and on Model Db was less than that generated by the SX2002 Spout-type Rainfall Simulator on the Model Da (Xu et al. 2006).

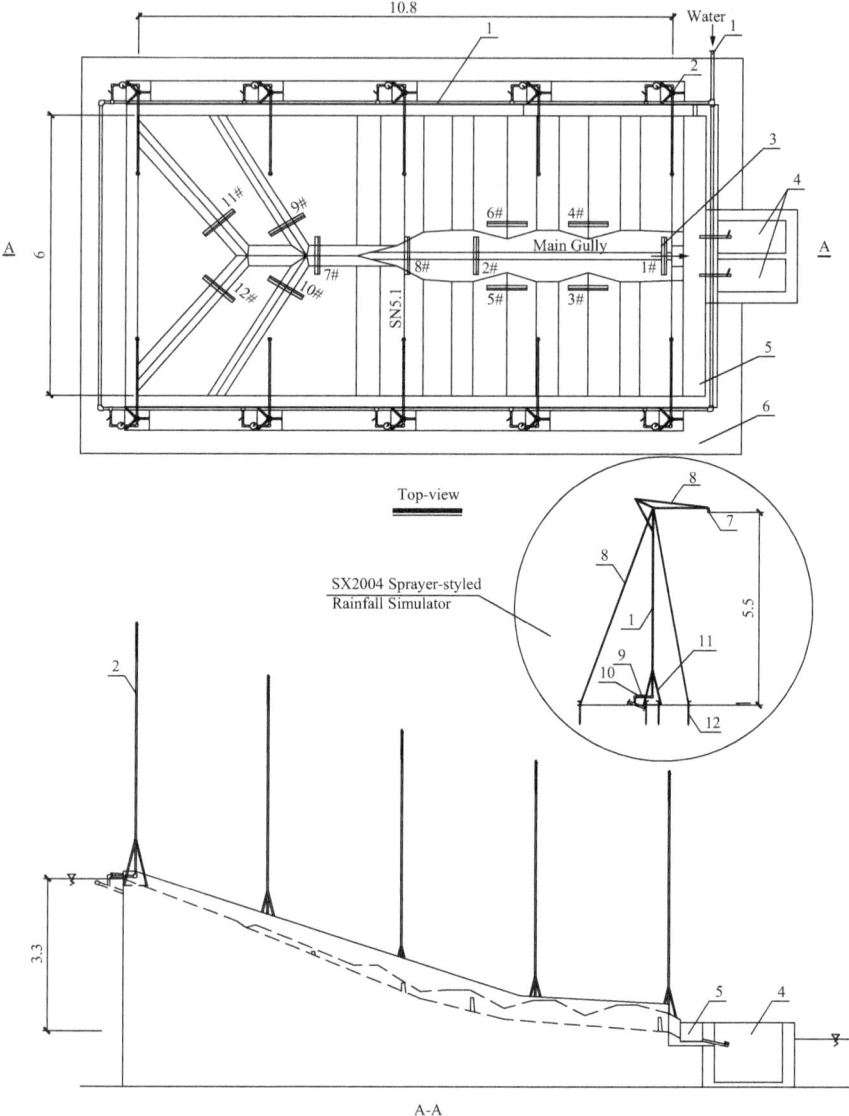

Fig. 2.2 Landscape simulator for the Model B in which rainfall simulation experiments were conducted. All units are in meters. Keys: 1. Water supply pipe; 2. Rainfall simulator; 3. Check dam; 4. Runoff container; 5. Collecting channel; 6. Path way; 7. Nozzle; 8. Guy rope; 9. Flow-meter; 10. Cut-off value; 11. Bracket; 12. Hawse-stick

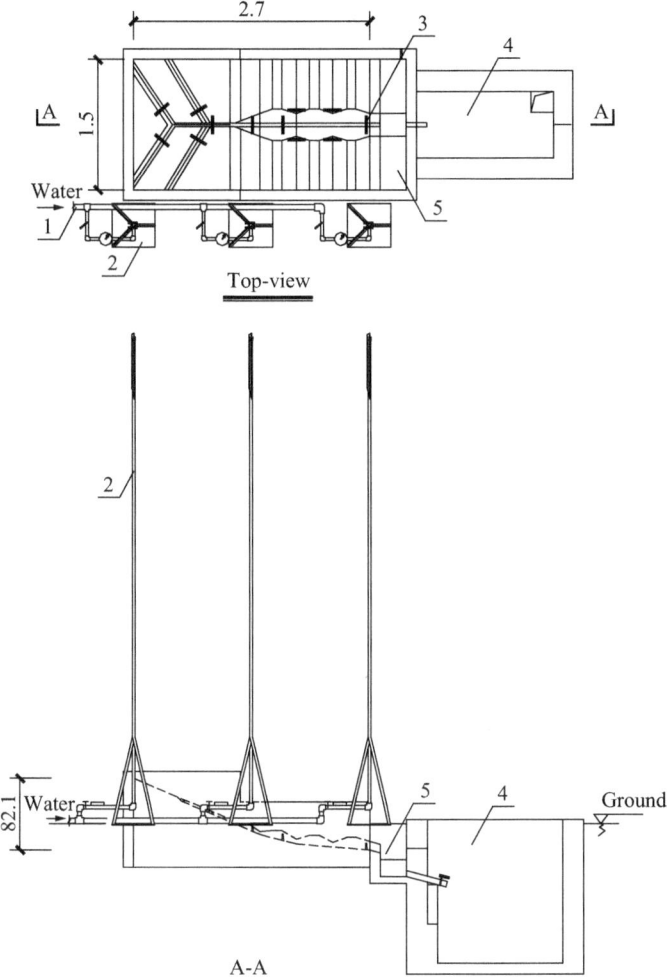

Fig. 2.3 Landscape simulator for the Model Db in which the rainfall simulation experiments were conducted. All units are in meters. Keys: 1. Water supply pipe; 2. Rainfall simulator; 3. Check dam; 4. Runoff container; 5. Collecting channel

Since the ratio R of the soil loss between the large scale model and the small scale models under rainfall events before check dam construction was determinate and pre-calibrated, and the erosive elements except for check dams, e.g., rainfall, and initial land cover, were unaltered. The ratio R was regarded as a constant after check dam construction in this study.

Brick walls were constructed around the model landscapes, with a knife edge of $60°$ facing away from the models on the upper rim of the models. The landscape was constructed using the loess soil collected from the Shunyi District, Beijing. The properties of the loess were close to those of the soil in the Yangdaogou Catchment.

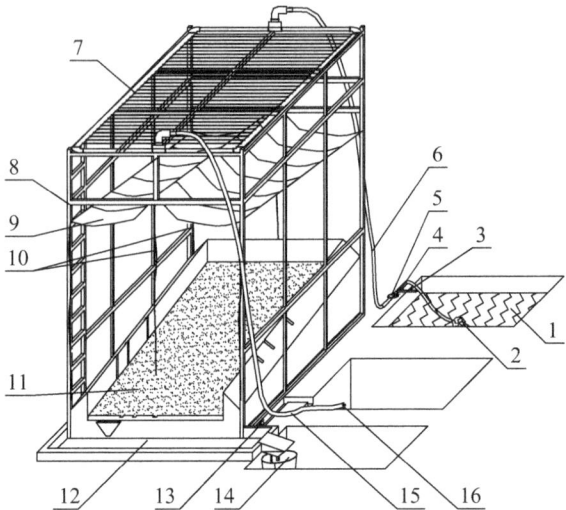

Fig. 2.4 Landscape simulator for the Model Da in which the rainfall simulation experiments were conducted. All units are in meters. Keys: 1. Constant head tank; 2. Immersed pump; 3. Branch pipe; 4. Valve; 5. Flow-meter; 6. Water supply pipe; 7. Rainfall generator; 8. Bracket; 9. Plastic screen; 10. Haul line; 11. Model landform; 12. Collecting ditch; 13. Drainage hole; 14. Bucket gauge; 15. Tail pipe; 16. Plug

The erodible material was passed through a 1 cm sieve to remove large clods, debris and grassroots. The basic properties of the loess are shown in Table 2.2. The 50% diameter of the soil particles d_{50} was 52.2 μm, and specific gravity, γ_s, was 2.56. However, in order to moderate the anti-erosion properties of the land cover in the Model Db, about 1/3 in volume of fine sand was mixed evenly with the loess, and some pre-designed rills were cultivated on the slopes.

An open-book-type watershed designed based on the landforms in the Yangdaogou Catchment in the 1960s, was used to represent the initial conditions for landscape development. To ensure a regular and original microtopography, the soil was prepared by hand patting to generate a 'smooth' roughness. A permeable trough base was used to allow free drainage of infiltrated water.

Table 2.2 Soil particle size distribution in the models

Particle diameter (mm)	*Cumulative weight (%)	Particle diameter (mm)	*Cumulative weight (%)
0.005	7.2	0.175	99.5
0.0125	22.2	0.35	99.8
0.0375	53.4	0.7	100.0
0.0875	96.5		

*The cumulative weight is the percentage of the total weight of particles with diameters less than the diameter to the total weight of the sand samples listed in the same column

2.4.3 Performance of the Landscape Simulator and Data Collection

Rainfall intensity was calibrated in advance of each rainfall event. The plot surface was covered with a tarpaulin. Rainfall was collected in a calibrated tank for the Model B or a small calibrated bucket for the Model Da and Model Db, and was volumetrically measured at 10-s intervals.

Soil water storage was very sensitive to rainfall (Peugeot et al. 1997). Thus, the significant effort was made to ensure equal initial soil moisture content before each experiment. To measure the initial soil moisture content, six soil samples were collected from the upper, middle and lower reaches of the model watershed surface, respectively, before each rainfall event. For Model B, an equal period of 24 h was maintained after each rainfall event. Before each rainfall-erosion event, a low-intensity rainfall was applied to the landform surface for a few minutes. For the Model Da and Model Db, before each rainfall-erosion event, atomized water drops from a portable nebulizer were applied to the land surface until the model landform was drenched but no runoff emerged.

The amount of soil loss for a downscaled watershed is typically estimated based on the measured amount of water discharged at outlets (Muttiah et al. 2005). During each rainfall event, the water-solid mixture was collected in a calibrated tank at the outlet of each plot. At 2 min or 1 min intervals, runoff samples were collected in 100 ml sampling bottles to determine the sediment concentration. Flow rates were synchronously measured using the label on the wall of the calibrated tank/bucket. After a rainfall event, the bed load deposited at the bottom of the tank was dried and weighed. The weight of the suspended load was calculated by the concentration determined gravimetrically, and the volume was then measured using the bucket gauge.

2.4.4 Executive Process

To validate the scale number of soil loss. Several groups of experiments were conducted to determine the model rain erosivity. Except that the initial landscape for the first rainfall event in a group of experiments was made by hand patting, the other subsequent landscapes were formed by the previous rainfall event. (1) Two groups of experiments were conducted for the Model B. First, 7 rainfall events, generated by the SX2004 Sprayer-styled Rainfall Simulator with intensities of about 1.60 mm/min and a duration of 20 min, were applied to the initial ground cover of the Model B without check dams. Then recovered the landform to the initial state, and gave 10 rainfalls same to the experiment above mentioned while 12 check dams were orderly constructed as shown in Table 2.3. (2) Two groups of experiments were conducted for the Model Da. First, 6 rainfall events, generated by the SX2002 Spout-type Rainfall Simulator, had intensities of about 1.65 mm/min and a duration of 10 min, were

Table 2.3 Comparison of the soil loss between the Model Da and Model B after the check dam system was constructed (R_{DB} = 2.50)

Run designation	Constructed dams	1	2	3	4	5	6	7	8	9	10
		Dam 1		Dam 2	Dam 3, 4, 5, 6	Dam 7	Dam 8, 9, 10, 11, 12				
Model B	Observed soil loss, S_B (kg)	40.10	138.10	73.00	44.70	48.90	47.10	36.70	47.20	61.30	40.60
	Soil loss converted in the prototype scale, S_{pB} (10^6 kg)	1.42	4.87	2.58	1.58	1.73	1.66	1.30	1.67	2.16	1.43
Model Db	Observed soil loss, S_D (kg)	5.04	6.72	4.48	1.51	1.40	2.11	1.65	2.73	1.64	1.55
	Soil loss free of rainfall difference, S_D^* (kg)	*	6.06	3.90	1.48	1.51	2.28	1.79	2.77	1.73	1.50
	Soil loss converted in the prototype scale, S_{pD} (10^6 kg)	*	5.46	3.52	1.33	1.36	2.06	1.61	2.50	1.56	1.35
	Observed rainfall intensity, I_D (mm/min)	2.70	1.71	1.74	1.64	1.56	1.56	1.56	1.61	1.58	1.65
	Soil loss error, e (%)	*	12.00	36.50	−15.50	−20.90	23.70	24.20	50.10	−28.00	−5.60

*Mis-operation

applied to the initial ground cover of the Model Da without check dams. The landform was then returned to the initial state. Then, 10 rainfall events, with the same properties as above mentioned were applied, while 12 check dams were orderly constructed (Table 2.3). (3) Two groups of experiments were performed for the Model Db. First, 6 rainfall events, generated by the SX2004 Sprayer-styled Rainfall Simulator with the intensities of about 2.45 mm/min and a duration of 10 min, were applied to the initial ground cover of the Model Db which was made of mixed soil and cultivated slopes before check dams were constructed. The landform was then returned to the initial state and 10 rainfall events, which had the same properties of the experiments above mentioned were applied, while 12 check dams were orderly constructed, as shown in Table 2.4.

To counteract the error of soil loss caused by the difference in the rainfall intensities. The rainfall intensity was strongly correlated with the soil loss in the experiments. Thus, a group of experiments was conducted on the Model Da to assess the relationship between the rainfall intensity and soil loss, and to identify a method for counteracting soil loss error caused by differences in rainfall intensities. In these experiments, no dam was built on the initial landform. The duration of each rainfall event was 10 min. However, the rainfall intensity in each rainfall event was different. The landform was returned to its initial state and pre-wetted before each rainfall event.

2.5 Results and Discussion

2.5.1 Counteraction of the Soil Loss Error Caused by Different Rainfall Intensity

Seven rainfall simulation experiments were conducted on the Model Da using the same initial landform and same initial water content. Table 2.5 presents the experimental results. For experiments under the following conditions of (1) with the same initial degree of saturation of bare land, (2) no large obstacle that blocks water, e.g., no check dam or pond, and (3) the same duration for each rainfall event, soil loss, S_D, was expressed as a power function of rainfall intensity, I_D:

$$S_D = 0.993 I_D^{1.98}, \quad r^2 = 0.94 \tag{2.5}$$

Similar experimental results were obtained by several experiments conducted on the Model Db without any dams. Accordingly, soil loss error caused by different rainfall intensities between the expected value and observed value could be counteracted as follows:

$$S_D^* = S_D \times (\bar{I}_D / I_D)^{1.98} \tag{2.6}$$

Table 2.4 Comparison of the soil loss between Model Db and Model B after the check dam system was constructed ($R_{DB} = 1.16$)

Run designation		1	2	3	4	5	6	7	8	9	10
Constructed dams		Dam 1		Dam 2	Dam 3, 4, 5, 6	Dam 7	Dam 8, 9, 10, 11, 12				
Model B	Observed soil loss, S_B (kg)	40.10	138.10	73.00	44.70	48.90	47.10	36.70	47.20	61.30	40.60
	Soil loss converted in the prototype scale, S_{pB} (10^6 kg)	1.42	4.87	2.58	1.58	1.73	1.66	1.30	1.67	2.16	1.43
Model Db	Observed soil loss, S_D (kg)	0.37	0.47	0.68	0.76	1.10	0.83	1.12	1.22	1.12	1.28
	Soil loss free of rainfall difference, S_D^* (kg)	0.45	0.47	0.69	0.80	1.16	1.10	1.02	1.14	0.87	1.12
	Soil loss converted in the prototype scale, S_{pD} (10^6 kg)	0.88	0.92	1.35	1.56	2.26	2.15	1.99	2.22	1.69	2.18
	Observed rainfall intensity, I_D (mm/min)	2.21	2.46	2.44	2.39	2.39	2.13	2.57	2.54	2.79	2.63
	Soil loss error, e (%)	−37.50	−81.20	−47.60	−0.90	31.20	29.40	53.60	33.10	−21.70	51.90

*Mis-operation

Table 2.5 Data from rainfall experiments conducted on the Model Da with the same initial landform but different rainfall intensities before the dam system was constructed

Sequence number	Experiment	Rainfall intensity, I_D(mm/min)	Soil loss, S_D(kg)	Regression analysis
1	C2-04-11-02-01(0.046)	2.24	4.66	$S_D =$
2	C2-04-10-26-01(0.044)	2.48	6.03	$0.993 I_D^{1.98}$
3	C2-04-10-24-01(0.042)	2.07	2.93	$r^2 = 0.94$
4	C2-04-10-23-01(0.040)	0.94	1.17	
5	C2-04-11-07-01(0.040)	1.71	2.99	
6	C2-04-11-22-01(0.038)	1.26	2.26	
7	C2-04-11-28-01(0.037)	0.54	0.21	

where S_D^* is the amount of soil loss free from the influence of the difference in rainfall intensities, and $\bar{I}_D$ is the expected rainfall intensity. In this study, $\bar{I}_D$ is the mean rainfall intensity in a group of rainfall experiments.

2.5.2 Calculation of Soil Loss Scale Number and Relative Ratio

Seven rainfall events were applied to the initial ground cover of the Model B without check dams. The rainfall-erosion in the fifth rainfall event ($I_B = 1.53$ mm/min, $S_B = 121.53$ kg), in which the initial soil density and moisture, and rainfall intensity were close to the condition in the experiments with check dams, was adopted as the index to calculate soil loss scale number of the Model B. In the same way, the amount of soil loss was 4.76 kg when the rainfall intensity was 1.65 mm/min in the Model Da.

Let R_{DB}, the relative ratio of erosion extent between the Model Db and Model B, be equal to 1.0. Then the amount of soil loss for the Model Db without any dams is shown as follows:

$$S_D = \frac{S_p}{\lambda_{LD}^3} = \frac{S_B \times \lambda_{LB}^3}{\lambda_{LD}^3} = \frac{121.53 \times 60^3}{240^3} = 1.90 \text{(kg)} \tag{2.7}$$

where λ_{LD} is the length scale number of the Model Db, and λ_{LB} is the length scale number of the Model B. After implementing a series of practices before the check dams were constructed, including increasing rainfall intensity, mixing fine sand as the land cover, and cultivating rills on slopes, the amount of soil loss in the Model Db was approximately 1.90 kg, 2.20 kg in practice, when the rainfall intensity was 2.45 mm/min.

The soil loss scale number of the Model B was calculated as follows according to its definition:

$$\lambda_{SB} = \frac{S_p}{S_B} = \frac{4.26 \times 10^6}{121.53} = 35,053 \qquad (2.8)$$

Calculated in the same manner, the soil loss scale numbers of the Model Da and Model Db were 9.01×10^5 and 1.95×10^6, respectively. The relative ratio of erosion extent between the Model Da (or Model Db) and Model B, R_{DB}, was calculated by solving Eqs. (2.3) and (2.4) simultaneously. Consequently, the relative ratio of the Model Da and Model Db to Model B was 2.5 and 1.16, respectively.

2.5.3   Validation of Soil Loss Scale Number

The rainfall-erosion data for the natural watershed, especially those after constructing dams, are so scarce that the reliability of the SSPM had to be verified by comparing the experimental results for the Model Da and Model Db with those for the Model B.

To validate the soil loss scale number as $R_{DB} \neq 1$. The Model Da was used to predict the amount of soil loss in the Model B when check dams were constructed in an orderly fashion. Thus, $R_{DB,}$ the relative ratio of the Model Da to Model B was 2.5 in these experiments. All soil losses in the model experiments were converted to prototype values by multiplying the soil loss scale number. The error of soil loss in the Model Da caused by the difference of the rainfall intensities was counteracted by Eq. (2.6). The relative differences in soil loss for the corresponding rainfall events between the Model B and Model Da, which represent the accuracy of the SSPM, were calculated using the following equation:

$$e = (S_{p_D} - S_{p_B}) \times 100\% / S_{p_B} \qquad (2.9)$$

where e is the relative differences of soil losses; S_{p_D} is the soil loss at the prototype scale that was transferred from the Model Da, $S_{p_D} = S_D \times \lambda_{SD}$; S_{p_B} is the soil loss at the prototype scale that was transferred from the Model B, $S_{p_B} = S_B \times \lambda_{SB}$. Consequently, the amount of soil loss in the Model Da was very close to that in the Model B, and the difference between the two models was less than 50% (Table 2.3), except that in the first rainfall event. The intensity of the first rainfall in the Model Da, 2.70 mm/min, was 63.9% more intense than the designed intensity resulting from human error, 1.65 mm/min.

To validate the soil loss scale number as $R_{DB} \approx 1$. The Model Db was used to predict the soil loss in the Model B when the check dams were built in an orderly fashion. The value of R_{DB} was 1.16 in these experiments. All soil losses in the model experiments were converted to the prototype values by multiplying the soil loss scale number. The soil loss error in the Model Db caused by the difference of rainfall intensities was counteracted using Eq. (2.6). Thus, the amount of soil loss in the Model Db was close to that in the Model B, and the difference between two models was less than 55% (Table 2.4), except for the second rainfall event with the soil loss error of 81.2%.

The precision in predicting the effect of sediment retained in the models was sufficiently good for the model experiments on soil conservation. It could be concluded that the experimental method can be used to predict the volume of soil loss in the small watershed of the Loess Plateau.

2.5.4 Qualitative Analysis of Erosion Depth

To compare the erosion trends in the model gullies with those in the prototype, the mean gully elevation in the Model Da and Model Db after each rainfall event was converted to that in the Model B using the scale-modification method. Notably, if the Model B is regarded as the simulated "prototype" and the Model Da as the downscaled model, then the length scale number is 4, and the soil erosion/deposition depth scale number is 1.6. However, if the Model B is considered as the simulated "prototype" and the Model Db as the downscaled model, the length scale number is 4, and the soil erosion/deposition depth scale number is 3.5.

The "SN5.1" is a branch gully without any check dam (Fig. 2.2). The mean elevation of this gully bed after each rainfall event is shown in Fig. 2.5. Elevations in the experiments with the Model Da and Model Db have been converted to the corresponding values at the Model B scale. The "prototype" deposition depth was strongly correlated with model deposition depths. Thus, we conclude that the geomorphological evolvement trends of both down-scaled models were similar.

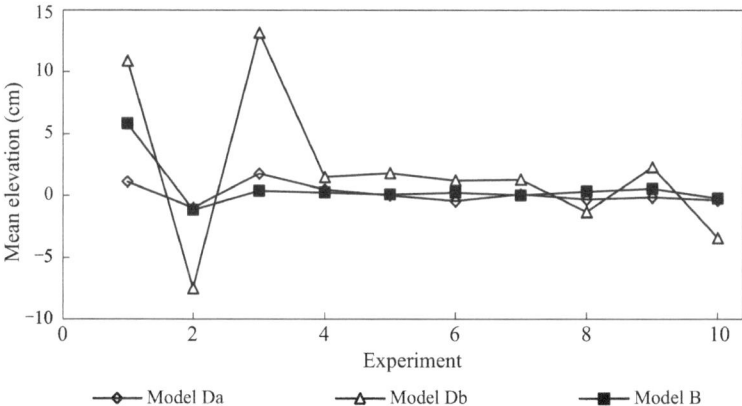

Fig. 2.5 Erosion-deposition in the gully SN5.1 where no dam was constructed. Depths in the Model Da and Model Db were converted to the corresponding values of the Model B scale

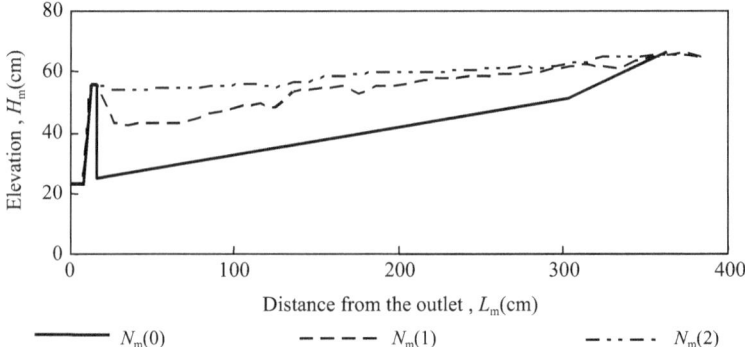

Fig. 2.6 The deposition process in the main gully at the upper reach of Dam 1

As illustrated in the regulations governing techniques for controlling erosion in gullies on the Loess Plateau (MWRC 2009), the design deposition life of a medium-sized check dam is 5–10 years, and that of a small-sized check dam is only 5 years. Field data demonstrated that the deposition occurring on the dam-land is extremely rapid when the hyper-concentrated flooding occurs. In this study, almost all check dams except Dam 1 were filled up after only one rainfall event. Even on dam-land of Dam 1, most of the deposition occurred during the first rainfall event, whereas the dam-land was filled up after 2 rainfall events (Fig. 2.6).

As the ratio R_B of the Model B is 6, meaning that the geomorphological evolvement rate after a single rainfall event in the Model B is about 6 times that in the prototype watershed in one year, the deposition lives of the check dams, except for Dam 1, were less than 6 years. The experimental results illustrate that deposition velocity on the dam-land simulated using model experiments was close to that in the prototype watershed on the Loess Plateau. Nevertheless, additional laboratory experiments for natural watersheds on the Loess Plateau are anticipated to further test the feasibility of the SSPM. This is a scientific challenge for future research.

2.6 Conclusions

When a check dam system is needed to retain soil on the Loess Plateau, it could be simulated using a downscaled model experiment. Data for the prototype watershed before dams are constructed should be utilized, including the rainfall, land cover and geological conditions of the prototype watershed. This study constructed the initial landform model in the laboratory, and applied simulated rainfall to the model. The simulated rainfall intensity was adjusted to conform to the ratio R. Finally, check dams were constructed according to the planned sequence and construction intervals of the check dams for each simulated rainfall event. The amount of soil loss after each rainfall event was transferred to the prototype one, which could predict the effect to retain the soil by check dams.

The SSPM is suitable for the design of the check dam system on the Loess Plateau, China. In this method, the ratio of the model geomorphological variables to the corresponding prototype remains constant after rainfall events. Consequently, the soil erosion processes in the prototype are reflected by the model experimental results. Experimental data in the serial experiments verify that the SSPM can predict soil conservation by a check dam system in a small watershed on the Loess Plateau, China.

References

Cai Q G, Wang G P, Cheng Y Z. 1998. The process and simulation of soil erosion for small watersheds on the Loess Plateau, China. Beijing: Science Press: 69–72, 135–137, 177, 189-190 (in Chinese).

Cerdà A, García-Fayos P. 2002. The influence of seed size and shape on their removal by water erosion. Catena, 48(4): 293–301.

Hancock G R, Willgoose G R. 2003. A qualitative and quantitative evaluation of experimental model catchment evolution. Hydrological Process, 17(12): 2347–2363.

Hancock G R, Willgoose G R. 2004. An experimental and computer simulation study of erosion on a mine tailings dam wall. Earth Surface Processes and Landforms, 29(4): 457–475.

Jiang D S, Zhou Q, Fan X K, et al. 1994. Simulated experiment on normal integral model of water regulating and sediment controlling for small watershed. Journal of soil and water conservation, 8(2): 25–30 (in Chinese).

Jin D S, Zhang O Y, Chen H, et al. 2003. Influence of base level lowering on sediment yield and drainage network development: an experimental analysis. Geographical Research, 22(5): 560–570 (in Chinese).

Li G Y. 2001. Construction of physical scale model for the yellow river. Yellow River, 23(12): 1–3,53 (in Chinese).

Muttiah R S, Harmel R D, Richardson C W. 2005. Discharge and sedimentation periodicities in small sized watersheds. Catena, 61(2): 241–255.

MWRC (Ministry of Water Resources, P.R.C). 2003. Programming for Check-dams in Loess Plateau (Technical Report): 47–48 (in Chinese).

MWRC (Ministry of Water Resources, P.R.C). 2009. Regulation of techniques for comprehensive control of soil erosion-Technique for erosion control of gullies. GB/T 16453.3-2008. Beijing: China Standard Press (in Chinese).

Peugeot C, Esteves M, Galle S, et al. 1997. Runoff generation processes: results and analysis of field data collected at the East Central Supersite of the HAPEX-Sahel experiment. Journal of Hydrology, 188–189(96): 179–202.

Shi H, Tian J L, Liu P L, et al. 1997a. Study on spatial distribution of erosion yield in a small watershed by simulation experiment. Research of Soil and Water Conservation, 4(2): 75–84,95 (in Chinese).

Shi H, Tian J L, Liu P L. 1997b. Study on relationship of slope-gully erosion in a small watershed by simulation experiment. Journal of Soil Erosion and Soil Water Conservation, 3(1): 30–33 (in Chinese).

Strabler A N. 1958. Dimensional analysis applied to fluvially eroded landforms. Bulletin of the Geological Society of America, 69(3): 279–300.

Tian Y H, Fu M S, Mu Z L, et al. 2003. Key technology to construct the check dam system. China Water Resources, (17): 59–61 (in Chinese).

Wang W Z, Jiao J Y. 1996. Rainfall and erosion sediment yield on the Loess Plateau and sediment transportation in the Yellow River Basin. Beijing: Science Press: 165, 221 (in Chinese).

Xu X Z, Zhang H W, Wang G Q, et al. 2006. A laboratory study on the relative stability of the check-dam system in the Loess Plateau, China. Land Degradation and Development, 17(6): 629–644.

Yuan J P, Jiang D S, Gan S. 2000a. Simulated experiment on normal integral model of different control degrees for small watershed. Journal of Natural Resources, 15(1): 91–96 (in Chinese).

Yuan J P, Lei T W, Jiang D S, et al. 2000b. Simulated experimental study on normalized integrated model for different degrees of erosion control for small watersheds. Transactions of the Chinese Society of Agricultural Engineering, 16(1): 22–25 (in Chinese).

Zhang H W. 1994. The study of the law of similarity for models of flood flows of the lower reach of the Yellow River. Beijing: Tsinghua University (in Chinese).

Zhang H W, Jiang E H, Bai Y M, et al. 1994. Similarity law for physical model of the hyperconcentrated flow in Yellow River. Zhengzhou: Henan Science and Technology Press: 80, 115, 156–162 (in Chinese).

Zhang Z G, Wang G P, Jia Z J, et al. 1995. Sediment source of Wangjiagou Gully in a well-controlled way. Science and Technology on Soil and Water Conservation in Shanxi, (2): 6–8 (in Chinese).

Zhang L P, Zhang M S. 2000. Research on runoff formation distortion coefficient in soil erosion experiment of a normal model of small drainage. Acta Pedologica Sinica, 37(4): 449–455 (in Chinese).

Chapter 3
A Conventional Experimental Technique: Rainfall Simulation

Abstract Rainfall simulation is widely used for the experimental study on soil and water conservation. This chapter reviews the characteristics, application, and prospect of rainfall simulator. The rainfall simulators could be divided into two main groups: the non-pressurized rainfall simulators, including the thread droppers and the needle droppers, and the pressurized water rainfall simulators, including the spout and the sprayer. The former could be hardly found at present due to its drawback that the minimum size of drop produced is far larger and more even than most natural rainfall and the simulator could not easily generate rains with the similar energy to the natural one. Rainfall simulation halls with automatic operation and observation systems are booming in recent years, but the basic simulators will not be completely replaced in virtue of the cheap price and convenient manipulations.

Keywords Rainfall simulator · Soil erosion · Research advance

3.1 Methods for Simulating the Rainfall

Rainfall is one of the major factors causing hydraulic erosions. However, natural rainfall experiments might require a considerable amount of time to complete. In contrast, not only various rainfall intensities may be obtained, but also shorter durations may be permitted in experiments under simulated rainfalls. Moreover, it is relatively easy to strictly control the experimental conditions and precisely monitor the processes of soil erosion in a rainfall simulation experiment. Especially for laboratory experiments, the rainfalls can only be obtained with rainfall simulators.

The uses of rainfall simulators have a long history in both laboratory and field investigations. With rainfall simulators, the United States and the Soviet Union began to model and monitor the processes of soil and water losses as early as the 1930s. China began to design and manufacture rainfall simulators in the 1960s. However, the technology has been developing rapidly in recent decades, and China has been at the forefront of designing and applying rainfall simulators.

To ensure the similarity between rainfall simulations is very important in the study of soil erosion. Nevertheless, differences may exist in raindrop characteristics between simulated rainfalls and natural events, such as raindrop size, velocity, and

© Science Press and Springer Nature Singapore Pte Ltd. 2020

X. Xu et al., *Experimental Erosion*,

https://doi.org/10.1007/978-981-15-3801-8_3

uniformity. Due to a lack of understanding of the effects of raindrops on soil erosion, the rainfall amount is generally used as the main parameter to simulate soil erosion under natural rainfall. However, numerous studies have shown that simulated rainfalls might result in obvious effects on soil erosion compared with natural rainfalls, even though rainfall amounts are the same in the above-mentioned rainfall events (Zhou et al. 1981).

Further studies have revealed that the effects of rainfall on soil loss varied with the kinetic energy of the raindrop (Watung et al. 1996; Zhou et al. 1981). Since soil erosion is triggered by the kinetic energy of falling raindrops, the energy is generally calculated and discussed as the main parameter during the simulated rainfall process. Wischmeier and Smith (1958) pointed out that the kinetic energy of the raindrop is the most suitable parameter to simulate the role of rainfall during soil erosion. This conclusion has been further confirmed by other researchers (e.g., Chen and Wang 1991; Park et al. 1983). In fact, the method of kinetic energy similarity is commonly used to simulate the natural-rainfall event. However, observing the rainfall energy is not as convenient as examining the rainfall intensity and drop size. Fortunately, there is a close relationship between rainfall kinetic energy and rainfall intensity (Xu et al. 2006a). Hence, the congruence in kinetic energy could be achieved by adjusting the rainfall intensity in rainfall simulations.

Many kinds of rainfall simulators have been developed and widely used in various experiments. Bubzenzer (1979) identified 63 rainfall simulators which are successfully applied in studies of soil and water conservation. Shelton et al. (1985) classified rainfall simulators into three forms, namely, hanging yarn, tubing tip, and nozzle, and the latter, however, is used almost exclusively for field plots because it has the capability to cover large areas. Wu and Xu (1995) further divided the rainfall simulators into four types. The first type is the thread dropper. Raindrops would drip from a hanging yarn to the ground along cotton threads or wool fibers. A distinct feature of this rainfall simulator is that the simulated raindrops are uniform and their initial falling velocities are zero. The raindrop sizes are determined by the diameter of the suspended thread instead of the water pressure of the water pipe. The second type is the needle dropper. Raindrops would drip from the ends of the needles to the ground. The simulated raindrops also fall to the ground with initial velocities of zero and their characteristics are comparable to raindrops from the thread droppers. The third type is the spout. The simulated raindrops are ejected at a certain initial velocity from small holes drilled on some parallel slender tubes, and then they are dropped to the ground. The raindrops have non-uniform diameters, and the rainfall intensity could be adjusted by different aperture sizes or water pressure. The fourth type is the nozzle/sprayer. Water drops are sprayed from the nozzle at a certain initial speed, then they would be dispersed into the air in various sizes, and finally, they would fall to the ground. The characteristics of this simulator are close to the features of the spout. Presently, the technique to simulate rainfall has been greatly improved around the world. Especially several rainfall simulation halls with automatic operation and observation systems are boosting in virtue of the combination of rainfall simulators and computers. Nevertheless, almost all the raindrop generators belong to the four types mentioned above (Wu and Xu 1995).

3.2 Rainfall Simulators with Thread Droppers or Needle Droppers

Various kinds of rainfall simulators may be designed to meet the demands of soil conservation experiments according to the mechanisms of the raindrop generator. As mentioned above, the spot and sprayer are frequently used all over the world. Table 3.1 shows the types and parameters of the rainfall simulators developed in recent years. It could be concluded that the hand-controlled basic simulators will continue to play an important role in the soil erosion experiments in virtue of convenient manipulation and low cost.

As mentioned before, the initial velocities of the raindrops generated from the thread droppers and needle droppers are zero. The raindrop falls on the ground along cotton or wool fibers from the water-supply pipe. One of the major drawbacks of these simulators is that the particle sizes of the simulated raindrops are larger than those of natural rainfalls. Moreover, different from the natural rainfall events, the distribution of drop sizes is almost uniform. Besides the cotton threads and wool fibers, the rainfall droppers may also be produced from fine glasses or brasses. Bowyer-Bower and Burt (1989) made a rainfall simulator with the Tygon tubing in a length of 15 mm, an internal diameter of 0.7 mm, and an external diameter of 2–3 mm. The former determined the rates of water drop formation while the latter controlled the sizes of the raindrops. The drop sizes were 2–3 mm or so.

The sizes of the raindrops formed by the needle-style simulators are close to those created by the thread droppers. However, relatively small particles would be generated if some measures were adopted, e.g., a wire mesh suspended beneath the drop-formers (Clarke and Walsh 2007). Gunn and Kinzer (1949) reported that by blowing a co-axial air stream directed vertically downwards over a single sized hypodermic needle and varying the rate of water flow, they could produce raindrops within an astonishing range of 0.25–10,000 mg.

By the end of the twentieth century, a large-scale thread-dropper rainfall simulator was built in the Institute of Geographic Science and Nature Resources Research, CAS. The simulator was equipped with a computer system to control the rainfall intensity, and an electric pendulum to crush the raindrops. In addition, a V-shaped groove was installed under each branch of pipe to discharge the overflow, which is very likely to generate large water drops. The rainfall intensity and raindrop size distribution of this kind of rainfall simulator could be adjusted by replacing the needles with different diameters. A high uniformity may also be achieved with the thread-dropper rainfall simulator.

The thread droppers and needle droppers have advantages of low-intensity threshold, high-intensity uniformity, and easy performance. However, in general, a relatively great height is required for the simulators mentioned above, since the sizes of the raindrops are large and their initial dropping velocities are close to zero. In fact, the raindrops would be spouted with a certain initial velocity if high-pressure water was used (Battany and Grismer 2000). In addition, the thread-droppers and needle-droppers are often inferior to the spouts and sprayers in achieving a similar

Table 3.1 Types and parameters of rainfall simulators manufactured in recent years

First author (completion time), and his/her research institute	Type of raindrop generator	Rainfall intensity (mm/min)	Fall height (m)	Characteristics
Wilson (1999), Commonwealth Scientific and Industrial Research Organisation, Australia	Multiple upward-spraying nozzles	0.5–2.5	3.0	The simulator was designed to be portable, able to produce a wide range of rainfall intensities, yield uniform and reproducible rain events, and represent the energy spectrum of natural raindrops
Esteves (2000), Laboratoire d'étude des Transferts en Hydrologie et Environnement, France	Multiple upward-spraying nozzles	1.1	6.6	The simulator was low cost, combined structure and easy to be used in field experiments
Battany (2000), University of California, Davis, USA	Downward pointing needles	0.3–1.5	3.5	The simulator framework and components were lightweight, readily available and easily manageable by one person. The device was easy to be used in small-scale laboratory experiments with greater rainfall distribution uniformity
Chen (2000), the Institute of Soil and Water Conservation, CAS & MWRC, China	Multiple downward-spraying nozzles	1.0–2.5	24.0	The simulator was appropriate for field experiments, which was easy to install, disassemble and transport with a simple structure
Wallbrink (2002), Cooperative Research Centre for Catchment Hydrology, Australia	Multiple upward-spraying nozzles	0.8–1.8	3.0	Two different sprinkler nozzles were used either singularly or in combination to produce three different rainfall intensities

(continued)

Table 3.1 (continued)

First author (completion time), and his/her research institute	Type of raindrop generator	Rainfall intensity (mm/min)	Fall height (m)	Characteristics
Sharpley (2003), USDA-ARS, USA	Multiple downward-spraying nozzles	1.2	2.5	The nozzles sprayed downward from an average height of 2.4 m, and then they produced drop-size distributions similar to natural rainfall. Rainfall uniformity was increased by rotating arms. Simulator energies were about 77% of those of natural rainfall
Blanquies (2003), California State University, USA	Multiple sideward-spraying nozzles	0.8	–	The structure of the simulator was built from aluminum, supporting the four-nozzle boom. Uniformity of rainfall was greater than 90% over the entire test plot. For one simulator the test plot was 3.56 m long and 1 m wide
Dang (2006), Soil and water conservation institute of Liaoning Province, China	Single sideward-spraying nozzles	0.5–3.0	10.0	The simulator with small size and light weight was easy to operate and suitable for field experiments
Xu (2006b), Tsinghua university, China	Downward-spraying spouts	1.0–4.0	4.0	The simulator was appropriate for laboratory experiments with good control-ability, stable performance and ease of operation. The rainfall area was 3.5 m × 2.5 m
Xu (2006b), Tsinghua university, China	Multiple downward-spraying nozzles	1.0–1.4	6.9	The large simulator was composed of 5 rainfall stands. It had been used in an experimental landform with 2.9 m of elevation difference and 70 m^2 of rainfall area

(continued)

Table 3.1 (continued)

First author (completion time), and his/her research institute	Type of raindrop generator	Rainfall intensity (mm/min)	Fall height (m)	Characteristics
Adams (2006), University of Massachusetts, USA	Multiple upward-spraying nozzles	0.6	5.0	The large simulator had 13 rainfall stands and the total application area of 1050 m^2, and it was easy to be used in field experiments
Clarke (2007), Cranfield University, UK	Downward pointing spout	3.3	1.4	Water drops were formed equally from each hole which equally spaced over the droplet box and a wire mesh was suspended below the drop-formers to vary the landing position of drops. It was easy to be used in the small-scale field research
Aksoy (2012), Istanbul Technical University, Turkey	Multiple downward-spraying nozzles	0.8–1.8	2.4	The simulator's compact structure fit a laboratory room. Its construction was simple and inexpensive. It could be operated by one person while additional persons were needed during runoff collection
Wildhaber (2012), University of Basel, Switzerland	Single downward-spraying nozzle	1.0	1.0–1.5	The simulator was a light weight, easy to handle irrigator and thus suitable for difficult to access alpine regions with low infrastructure, steep slopes and uneven terrain
Salem (2014), Polytechnic University of Madrid, Spain	Multiple sideward-spraying nozzles	0.6–1.9	2.3	The simulator was an auto-controlled laboratory rainfall simulator that could obtain variable rainfall intensities with drop sizes similar to natural rain
Lora (2016), University of Padua, Italy	Multiple sideward-spraying nozzles	0.8–2.5	3.7	Limited impact energy on the soil in order to avoid surface erosion. A flexible manageability of the simulator

(continued)

Table 3.1 (continued)

First author (completion time), and his/her research institute	Type of raindrop generator	Rainfall intensity (mm/min)	Fall height (m)	Characteristics
Al (2017), University Libanaise, France	Single downward-spraying nozzle	2.0	1.6	The simulator was designed to homogeneously spray an elementary area of 1 m². It is lightweight and it could be easily operated by on person and to reduce implementation time to less than an hour
Kesgin (2018), Yildiz Technical University, Turkey	Multiple downward-spraying nozzles	0.4–4.4	6.5	The simulator could produce different rainfall rates to be able to obtain desired hyetograph. Spatial uniformity of produced rainfall was ranging from 90.24 to 92.74%
Vergni (2018), University of Perugia, Italy	Multiple downward-spray nozzles	0.3–2.3	2.8	The simulator were usually employed in field experiments, both single and multi-nozzle operation were feasible, events were simultaneously reproduced in two identical plots, and the runoff volume coming from the micro-plots and the water volume falling outside the plots were both collected and conveyed to separate outlets

kinetic energy to natural rainfall, and the droppers are susceptible to clogging especially if the supplied water is turbid. The thread-droppers and needle-droppers may be used in a relatively small experimental area. Presently, they could be hardly found although they had been used in the early studies.

In practice, a mesh screen is often suspended under the thread droppers to produce droplets with small sizes close to those of natural rainfalls. Nevertheless, due to surface tension, water may be accumulated in some areas of the screen and then outsized drops might be formed. In addition, in the early phases, rainfalls from the thread-droppers or needle-droppers are unstable; shortly after the power source is turned off, some "redundant" raindrops may be produced with large sizes. Hence, rainfalls in the two stages mentioned above should be excluded in a rainfall simulation experiment in order to ensure the accuracy of the experimental results.

3.3   Rainfall Simulators with Spouts or Sprayers

Since the 1960s, rainfall simulators have been developing from non-pressurized rainfall simulators into sprayer/spout rainfall simulators. The principle behind the use of pressurized water is that the drops sprayed out from the nozzle have an initial velocity which may be sufficient for the drops to reach their terminal velocity at a considerably less falling height than drops falling from the skies. This reduction in necessary falling height is a notable advantage for pressurized simulators over non-pressurized simulators which rely on gravity and free fall of drops to attain terminal velocity. The pressurized water rainfall simulators could be divided into two main groups: the spout rainfall simulators and sprayer rainfall simulators.

The spout rainfall simulators are widely used in experimental studies of soil erosion. A relatively short distance is needed from the simulator to the model surface because the water ejects from the nozzle with an initial speed that is large enough to make water drops break and scatter. Hence, the characteristics of the simulated rainfall are closer to those of the natural rainfall. In order to study the interflow in soil, a scientist from the former Soviet Union designed a kind of simulator with fine nozzle clusters. Nevertheless, the major drawback of this rainfall simulator is that the area covered by this simulator is small, and the height of the simulator is not sufficient to make raindrops attain their terminal velocities at which natural raindrops with the same sizes impact the ground (Zhao and Li 1989). A small spout-type rainfall simulator was also developed by the Institute of Geographic Sciences and Natural Resources Research, Chinese Academy of Sciences, which could be used in infiltration experiments with areas of 1 m². Some of the rainfall simulators designed by the authors of the monograph are shown in Figs. 3.1, 3.2, 3.3 and 3.4. The SX2002 Spout-type Rainfall Simulator designed by Xu et al. (2006b) might be used in an experimental plot measured 2.5 m × 3.5 m. A high initial spouting speed together with a falling height of 4 m is enough to make the raindrops reach their terminal velocities. The uniformity coefficients of the rainfall intensities for the simulators exceeded 80%. Moreover, the rainfall intensity could be adjusted by controlling the

Fig. 3.1 A rainfall simulator with downward-spraying spouts (Xu et al. 2006b). The simulator was appropriate for laboratory experiments with good controllability, stable performance and ease of operation. The rainfall area was 3.5 m × 2.5 m

Fig. 3.2 A spout type rainfall simulator with upward orifices (Xue 2007). Raindrops spouted obliquely upwards, hit the stream breaker, and then spread around the ground

Fig. 3.3 Multiple downward-spraying nozzles (Xu et al. 2006b). The simulated rainfall covered an experimental landform with an elevation difference of 2.9 m and an area of 70 m^2

Fig. 3.4 A rainfall simulator with multiple downward-spraying nozzles (Xu et al. 2015)

pressure of supply water in the range of 1.0–4.0 mm/min. The merit was better than the spray-type simulator. Based on this kind of simulator, Xue (2007) designed a spout-type rainfall simulator with upward orifices. Raindrops spouted obliquely upwards, hit the stream breaker, a horizontal plate hung above the pipe network, and then spread around the ground. Because the nozzles pointed upwards, the rainwater might automatically retreat when rainfall was stopped, and thus blocking was not common. The uniformity of the rainfall simulator was increased because the stream breaker made the raindrops scattered.

The sprayer rainfall simulator is one of the most widely used devices in soil-erosion experiments. On the basis of nozzle direction, the spray-type rainfall simulator consists of nozzles pointing upwards, nozzles pointing downwards, and nozzles pointing sideward. There is a rotatable sprinkler in each nozzle pointing upwards or downwards, which may produce an arc of water, spread the water current in a large area, and break the current into pieces. In comparison, an oblique board is installed in the nozzle pointing sideward. Water rushes to the oblique board, and then changes the direction and horizontally run out of the sprayer to form scattered raindrops in a large area. According to the scattering mechanism, spray-type rainfall simulators may be classified as one of the rotation type simulators, where a rotating helical blade is used to spread the rainfalls, and as one of the aeration type, in which water is atomized to improve the uniformity coefficients. In addition, the spray-type rainfall simulators may also include movable and stationary type based on the movement style of the nozzle.

As mentioned before, most simulators currently used in infiltration experiments are in the sprayer forms. Meyer and McCune (1958) invented a rainfall simulator with a downward sprayer alternately moving to and fro on a slide bar. The simulator achieved good uniformity of rainfall intensity. The rainfall simulator designed by Esteves et al. (2000) included spraying nozzles mounted on the top of the pipe at a height of 6.5 m. Water was jetted to a height of approximately 7.5–8 m. Under the water pressure of 41.4 kPa, the mean drop diameter was 2.4 mm, the calculated kinetic energy was 23.5 J/(m^2 mm), and the rainfall intensity was 1.1 mm/min. Ye et al. (2001) carried out a field experiment to analyze erosion due to railway construction on representative slopes. Wu et al. (2003) designed a sprayer rainfall simulator to investigate the benefits and characteristics of grass-shrub vegetation for reducing soil erosion in the Loess Hill Ravine Region. The sprayer rainfall simulators are also applied to test the soil loss of compressed loess roadbed (Shen et al. 2003) and the infiltration under the conditions of conservation tillage (Wang et al. 2000). The simulator introduced by Sharpley and Kleinman (2003) was trailer-mounted, with 10 rotating booms (each 7.6 m long) radiating from a central stem, which rotated at about 4 r/min. The nozzles sprayed downward from an average height of 2.4 m, applied the rainfall intensity of 70 mm/h and produced drop-size distributions similar to natural rainfall. In recent years, no major revision has been made on the raindrop generators, even though the sprayer rainfall simulators have been used in experiments under various operating conditions. Many of them are designed for field studies, which are easy to handle and suitable for the areas difficult to access (e.g., Al et al. 2017; Lora et al. 2016; Wildhaber et al. 2012). Some of the field simulators are at very large scales. The simulator designed by Adams and Elliott (2006) has 13 rainfall stands at 9 m spacing, and the total application area under the sprinkler stands was greater than the catchment area of the plot (approximately 1050 m^2).

Compared with the sprayer-styled rainfall simulator, the coverage area and uniform coefficient of spout-type rainfall simulator are less influenced by water supply pressure, and the adjustable range of the rainfall intensity is relatively large. However, similar to the needle droppers, the spouts also have to overcome the unsteady effects of pre-rainfall in the beginning and residual water at the end in an experiment. Besides, a spray-type or sprout-type rainfall simulator may be composed of several independent rainfall units. The flexible combination of the units can overcome the difficulties of the great difference in the elevation of the underlying surface and the large scale of coverage area of simulated rainfall. In practice, better than the spout-type rainfall simulator, a spray-type simulator may be employed with moveable nozzles to make rainfall distribution more uniform, or expand the area covered by the rainfall in the case of fewer sprinklers.

3.4 Automated Rainfall-Simulation Hall

With the development of soil erosion research, many rainfall-simulation halls equipped with comprehensive advanced facilities have recently emerged. The world's largest rainfall simulation hall was built in 1974 by the National Research Institute for Earth Science and Disaster Prevention, Japan, and it still undertakes the experimental study of soil and water conservation until present (NIED 2015). The hall has an effective rainfall simulation area of 3168 m^2, which measures 44 m × 72 m, and a rainfall intensity in the range between 0.25 and 3.30 mm/min. Drops sprayed out of the combined nozzle rainfall simulator with automatic control have an initial fall velocity, and the height of the nozzles is 16 m above the ground which is far enough for raindrops to reach their terminal velocity. Several academic institutes in China, e.g., Institute of Soil and Water Conservation—Chinese Academy of Sciences & Ministry of Water Resources (ISWC), Institute of Geographic Sciences and Natural Resources Research—Chinese Academy of Sciences, Heilongjiang Institute of Soil and Water Conservation, and Xi'an University of Technology, have constructed highly automated rainfall-simulation halls. The hall built by the ISWC is the second largest but the most advanced in the world. It has an effective application area of 1296 m^2, which is equipped with a series of experimental devices including the downward-spraying or sideward-spraying rainfall simulators, and stationary or moveable hydraulic lifting and descending soil-bins. Other advanced technologies also have been employed, consisting of the computer system automatically controlling the rainfall characteristics and the observation device dynamically monitoring erosion processes on the slope.

Large-scale model experiments for soil and water conservation could be conducted in the rainfall simulation hall because in the hall, the rainfall area is so large, and the distribution of the simulated rainfall energy closely approximates natural rainfall. The result of soil erosion research will be more reliable and efficient in virtue of the progressive equipment such as the instrument dynamically measuring soil loss and facility automatically controlling rainfall. Nevertheless, the cost of construction and maintenance is expensive for the large-scale indoor rainfall simulator.

3.5 Conclusions

Rainfall simulation plays an important role in the study of soil erosion. The rainfall simulators can be categorized into four groups according to the formation of raindrops: the thread dropper, the needle dropper, the spout, and the sprayer. The sprayer is widely used in soil conservation experiments. Moreover, rainfall simulation halls with automatic operation and observation systems are booming in recent years, but the basic simulators will not be completely replaced in virtue of the cheap prices and convenient manipulations.

References

Adams R, Elliott S. 2006. Physically based modelling of sediment generation and transport under a large rainfall simulator. Hydrological Processes, 20(11): 2253–2270.

Aksoy H, Unal N E, Cokgor S, et al. 2012. A rainfall simulator for laboratory-scale assessment of rainfall-runoff-sediment transport processes over a two-dimensional flume. Catena, 98: 63–72.

Al A S, Bonhomme C, Dubois P, et al. 2017. Investigation of the wash-off process using an innovative portable rainfall simulator allowing continuous monitoring of flow and turbidity at the urban surface outlet. Science of the Total Environment, 609: 17–26.

Battany M C, Grismer M E. 2000. Development of a portable field rainfall simulator for use in hillside vineyard runoff and erosion studies. Hydrological Processes, 14(6): 1119–1129.

Blanquies J, Scharff M, Hallock B. 2003. The design and construction of a rainfall simulator. International Erosion Control Association (IECA), 34th Annual Conference and Expo. Las Vegas: 044.

Bowyer-Bower T A S, Burt T P. 1989. Rainfall simulators for investigating soil response to rainfall. Soil Technology, 2(1): 1–16.

Bubzenzer G D. 1979. Inventory of rainfall simulators. Proceedings of the rainfall simulator workshop, Tucson, AZ: 120–130.

Chen W L, Tang K L. 2000. A new SR style field artificial rainfall simulator. Research of Soil and Water Conservation, 7(4): 106–110 (in Chinese).

Chen W L, Wang Z L. 1991. The trial research on the behaviours of artificial rainfall by Simulation. Bulletin of Soil and Water Conservation, 11(2): 55–62 (in Chinese).

Clarke M A, Walsh R P D. 2007. A portable rainfall simulator for field assessment of splash and slopewash in remote locations. Earth Surface Processes and Landforms, 32(13): 2052–2069.

Dang F J, Ge S F, Zheng J. 2006. A circular on research of rainfall simulator-model DQSY. Science of Soil and Water Conservation, 4(5): 99–102 (in Chinese).

Esteves M, Planchon O, Lapetite J M, et al. 2000. The 'EMIRE' large rainfall simulator: design and field testing. Earth Surface Processes and Landforms, 25(7): 681–690.

Gunn R, Kinzer G D. 1949. The terminal velocity of fall for water droplets in stagnant air. Journal of Meteorology, 6(4): 243–248.

Kesgin E, Dogan A, Agaccioglu H. 2018. Rainfall simulator for investigating sports field drainage processes. Measurement, 125: 360–370.

Lora M, Camporese M, Salandin P. 2016. Design and performance of a nozzle-type rainfall simulator for landslide triggering experiments. Catena, 140: 77–89.

Meyer L D, McCune D L. 1958. Rainfall simulator for runoff plot. Agricultural Engineering, 39(10): 644–648.

NIED (National Research Institute for Earth Science and Disaster Prevention), JAPAN. 2015. Large-scale rainfall simulator—the globe's No. 1 scale and capability. [2018-12-01]. http://www.bosai. go.jp/e/facilities/rainfall.html.

Park S W, Mitchell J K, Bubenzern G D. 1983. Rainfall characteristics and their relation to splash erosion. Transactions of the ASAE, 26(3): 0795–0804.

Salem H M, Valero C, Muñoz M Á, et al. 2014. Effect of reservoir tillage on rainwater harvesting and soil erosion control under a developed rainfall simulator. Catena, 113: 353–362.

Sharpley A, Kleinman P. 2003. National phosphorus runoff project: Pennsylvania - rainfall simulator and plot scale comparison. Journal of Environmental Quality, 32(1): 2172–2179.

Shelton C H, Von Bernuth R D, Rajbhandari S P. 1985. A continuous-application rainfall simulator. Transactions of the ASAE, 28(4): 1115–1119.

Shen B, Zheng N X, Tian W P. 2003. The research on the rainfall erode of the compress roadbed slope-surface. Journal of Chongqing Jiaotong University, 22(4): 64–67 (in Chinese)

Vergni L, Todisco F, Vinci A. 2018. Setup and calibration of the rainfall simulator of the Masse experimental station for soil erosion studies. Catena, 167: 448–455.

Wallbrink P J, Croke J. 2002. A combined rainfall simulator and tracer approach to assess the role of Best Management Practices in minimising sediment redistribution and loss in forests after harvesting. Forest Ecology and Management, 170(1–3): 217–232.

Wang X Y, Gao H W, Du B, et al. 2000. Conservation tillage effect on runoff and infiltration under simulated rainfall. Bulletin of Soil and Water Conservation, 20(3): 23–25,62 (in Chinese).

Watung R L, Sutherland R A, El-Swaify S A. 1996. Influence of rainfall energy flux density and antecedent soil moisture content on splash transport and aggregate enrichment ratios for a Hawaiian Oxisol. Soil Technology, 9(4): 251–272.

Wildhaber Y S, Bänninger D, Burri K, et al. 2012. Evaluation and application of a portable rainfall simulator on subalpine grassland. Catena, 91(3): 56–62.

Wilson C J, Wallbrink P J, Murray A S. 1999. Minimum impact logging systems in forest erosion. CSIRO Land and Water, Canberra, Technical Report 19/99, [2018-12-01]. http://www.clw.csiro. au/publications/technical99/tr19-99.pdf.

Wischmeier W H, Smith D D. 1958. Rainfall energy and its relation to soil loss. Transactions, American Geophysical Union, 39(2): 285–291.

Wu C W, Xu N J. 1995. The experiment on properties of rainfall-simulator of swayed sprinkler. Journal of Nanchang University, 17(1): 58–66 (in Chinese).

Wu Q X, Zhao H Y, Han B. 2003. Benefit and characteristics of grass-shrub vegetation for reducing soil erosion in loess hilly region. Acta Agrestia Sinica, 11(1): 23–26 (in Chinese).

Xu X Z, Liu Z Y, Xiao P Q, et al. 2015. Gravity erosion on the steep loess slope: Behavior, trigger and sensitivity. Catena, 135: 231–239.

Xu X Z, Liu D Q, Zhang H W, et al. 2006a. Laboratory rainfall simulation with controlled rainfall intensity and drainage. Journal of Beijing Forestry University, 28(5): 52–58 (in Chinese).

Xu X Z, Zhang H W, Dong Z D. 2006b. Experimental study on SX2002 Pipe-Network-Type precipitation simulation apparatus. Soil and Water Conservation in China, (4): 8–10,52 (in Chinese).

Xue Y N. 2007. Design of indoor simulation test system based on efficiency analysis of rainwater resource utilization. Dalian: Dalian University of Technology: 16–22 (in Chinese).

Ye C L, Xu Z Y, Yang C Y. 2001. Study on water and soil loss during procession of constructing qinhuangdao-shenyang special line for passengers train. Journal of Soil and Water Conservation, 15(2): 9–13 (in Chinese).

Zhao Z J, Li G Y. 1989. Review and prospect on equipments and methods for rainfall stimulating. Soil and Water Conservation in China, (5): 30–33 (in Chinese).

Zhou P H, Dou B Z, Sun Q F, et al. 1981. Preliminary experimental study of rainfall energy. Bulletin of Soil and Water Conservation, (1): 51–61 (in Chinese).

Chapter 4
An Innovative Measurement Instrument: Topography Meter

Abstract The measurement of failure mass is very difficult because gravity erosion usually occurs randomly and it combines with hydraulic erosion. Here, we present a novel structured-light 3D surface measuring apparatus, the topography meter, which could quantitatively measure the time-variable gravity erosion on the steep loess slopes. With the topography meter, a 3D geometric shape of the target surface could be digitally reconstructed, and then, the slope parameters, including the volume, projected area, and gradient distribution, could be obtained. By comparing the slope geometries in the moments before and after the erosion incident on the snapshot images at a particular time, we could obtain the volume of gravity erosion and many other erosion data. A series of calibration tests were conducted and the results showed that the accuracy of this technique was high and sufficient for exploring the mechanism of slope erosion.

Keywords Natural hazard · Gravity erosion · Topography meter · Erosion volume

4.1 Quantitative Monitoring of Gravity Erosion

Applicable and precise instruments are vital to obtain credible data in experiments. With the development of photoelectric, ultrasonic, and image processing technologies (Meng and Janssen 2015; Thomas et al. 2014), noncontact observation instruments with high efficiency and accuracy (e.g., digital photogrammetry and laser method) have been invented and used in the survey of the terrestrial landform in laboratories. However, strong sunshine and wind will present great difficulties in making precise topographic measurements in the field, especially for the laser method. Determining the best method to manage problems appearing in terrain evolution experiments is an important subject of survey engineering that must be addressed.

Terrestrial laser scanning is a technology that allows the digital acquisition of real objects, reproducing them in 3D space and digital form employing a cloud of points (Du et al. 2013; Casula et al. 2010; Vondrak et al. 2006). Compared to conventional methods such as triangulation, field and office time of operation was reduced using laser scanning and GPS. Applications where safety may be an issue, such as providing accurate measurements on a landslide or debris flow area, would

© Science Press and Springer Nature Singapore Pte Ltd. 2020
X. Xu et al., *Experimental Erosion*,
https://doi.org/10.1007/978-981-15-3801-8_4

benefit most from the strengths of this technology (Du and Teng 2007; Abellán et al. 2006). Nevertheless, while a terrestrial laser scanner was used to quantify rills on an angle of repose slope, an even greater density of points was needed to capture sufficient rill morphology (Scherer and Lerma 2009; Hancock et al. 2008). Moreover, it is very difficult to monitor the time-variable process of an individual mass failure with these techniques because of the randomness and suddenness of such an event.

Photography was invented in 1839 by a Frenchman, Louis Daguerre, and a year later terrestrial photos were first used for surveying (Wolf 2002). Compared to terrestrial laser scanning, the philosophy of terrestrial photos captured the geometry and visualized it simultaneously; the coordinates of relevant object-describing points were determined by the intersection of an image ray and the predefined primitive (Scherer and Lerma 2009; Ohnishi et al. 2006). Digital photogrammetry was applied for measuring erosion rates on complex-shaped soil surfaces under laboratory rainfall conditions (Rieke-Zapp and Nearing 2005). However, the images of the object taken with a camera had to be geo-referenced as a preparatory step, and an intelligent tacheometry was used for online and on-site measurements (Scherer and Lerma 2009; Rieke-Zapp and Nearing 2005). Structured-light 3D surface imaging techniques were used to compute the 3D model of a face by projecting a simple colored stripe pattern onto the face, and the depth information was then calculated by considering the distortion of the stripes in the face caused by its shape (Geng 2011; Wang et al. 2010; Fechteler and Eisert 2009; Peng 2007; Zhang 2005). If the target was moving, single-shot techniques had to be used to acquire an instant snapshot 3D surface image of the 3D object at a particular time. Nevertheless, to our knowledge, no such techniques have been applied to the 3D object in dynamic motion in literature before.

Recently, a structured-light 3D surface-measuring instrument, the MX-2010-G topography meter, was designed and manufactured to observe the slope behavior under rainfall simulation. Evaluation tests showed that the errors among the volumes observed by the topography meter and those of the conventional instruments were within 10% (Xu et al. 2015). However, the above-mentioned paper was focused on data processing and calibration testing. It did not give detailed information on the structure of the 3D surface measuring instrument. In addition, the definition of laser lines recorded by the camera had a high requirement for the ambient light. If the ambient lighting was sufficiently strong, the contrast of the laser footprints would decrease on the sloping terrain, and the image would become vague. Whether it is possible to use the topography meter for conducting site-specific surveys of landslides has become an urgent question to be resolved.

Gravity erosions, the mass failures on steep slopes that are triggered by self-weight, contrast with other soil erosions requiring physical impetus of wind or water, and were among the most important natural hazards in mountainous regions (Hergarten 2012; Gokceoglu and Sezer 2009). Gravity erosion tends to happen as an episodic event in which large sections of the steep slope fail. However, to monitor the time-variable process of an individual mass failure is very difficult because of the randomness and suddenness of such an event. It is rather difficult to conduct a field survey of soil and water conservation on the Loess Plateau of China. During the natural rainfall events, the site-specific observation of gravity erosion is almost impossible, because we are unable to predict when and where the gravity erosion will occur. Hence, what we could do is to simulate the landslide processes by employing a conceptual slope under rainfall simulation. However, the representative catchments on the Chinese Loess Plateau for scientific research are generally far from urban areas, lack water and electricity, and have deep ditches, steep slopes, inconvenient transportation, and high-strength winds. The special situation poses another severe challenge to the site survey. Here, we designed a removable house that could be assembled in the field to conduct site-specific tests based on the terrain of the Loess Plateau. The room provided the same conditions for simulations and observations as those in the laboratory. We also presented a detailed illustration of the MX-2010-G topography meter to dynamically monitor the slope behavior under rainfall simulation. The design optimization and fabrication methods of the topography meter and the laboratory would be expected to be useful for congeneric device applications.

4.2 Experimental Setup and Methods

The topography meter consists of the following conventional components: a camera, a laser source, and a positioning device. Figure 4.1 shows a representative scheme for quantitative monitoring of gravity erosion with the structured-light 3D surface

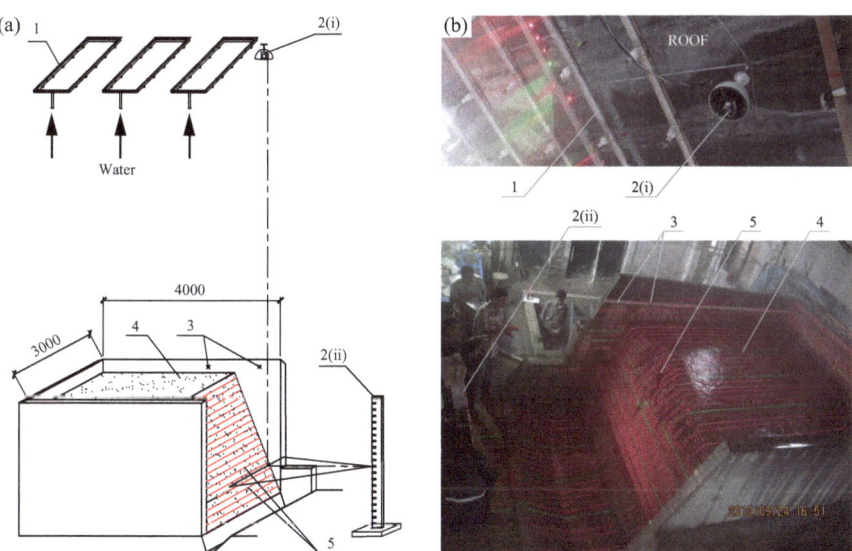

Fig. 4.1 A representative scheme for quantitative monitoring of gravity erosions with the structured-light 3D surface measuring technique. **a** Blue print of the topography meter measurement system; **b** Picture of an experimental site. Keys: 1. Rainfall simulator; 2. Topography meter (i—Camera with a collimator, ii—Laser source); 3. Positioning marks; 4. Model slope; 5. Equidistant horizontal projections

measuring apparatus. The horizontal stripe pattern with a 3.0 cm contour interval is generated by a laser source and is recorded by the camera with the sighting direction perpendicular to the laser planes. Positioning marks are placed on fixed positions with the same height. As the slope terrain deforms over time, the erosion process is recorded by video, and then imported into the computer to acquire a snapshot image at a particular time sequence. Given depth in ArcGIS, the 3D geometric shape of the target surface can be computed accurately. By comparing the slope geometries in the moments before and after the erosion incident, we could obtain the volume of gravity erosion and many other erosion data. In this study, the topography meter covers a monitoring range of 3.0 m × 2.0 m, and the SONY HDR-XR550E video camera has a resolution of 6.63 Megapixel.

In order to testify the validity and accuracy of the topography meter on detailed gravity erosions, calibration tests were conducted on conceptual slopes. The topographical characteristics were initially measured by the conventional instruments, e.g., steel rule and level instrument, and then surveyed with the topography meter. The tests consisted of two stages, as shown in Figs. 4.2 and 4.3. In the first stage, a wooden brae was measured with the topography meter, and then the lengths of brae edges covered with the laser footprints were gauged with a steel rule to calculate the volume of the brae. Subsequently, the brae was placed in the other gradients, and also observed by the topography meter and the steel rule respectively. In the second stage,

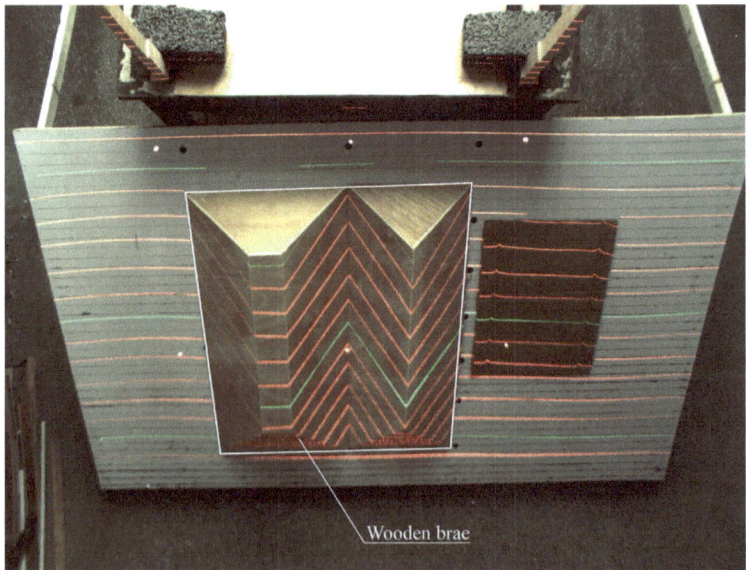

Fig. 4.2 A laboratory calibration test for measurement of the slope volume. The wooden brae was placed in various slope gradients, and was investigated by the topography meter. Then the bulk of the brae covered by the laser beam was elaborately gauged by a steel rule

Fig. 4.3 A field calibration test for measurement of hillside coordinates. The terrain of a natural hillside was investigated with the topography meter. Then grid points of the hillside were measured with a steel rule and a level instrument to get the topography map. Keys: 1. Camera with a collimator; 2. Landform covered with laser beams; 3. Positioning marks; 4. Laser source; 5. Level instrument

a natural hillside was tested with the topography meter, and then grid points of the hillside were measured with a steel rule and level instrument to get the topography map.

The topography meter was applied to real gravity erosion experiments. In the Joint Laboratory for Soil Erosion of Dalian University of Technology and Tsinghua University, a series of gully bank collapse experiments under closely controlled conditions were conducted. The landscape simulator consisted of a rainfall simulator and a flume containing the slope model, as shown in Fig. 4.1. The simulator, which covered an area of 3.0 m by 3.0 m, was formed of a framework of steel pipes with 30 sprinkling nozzles evenly arranged 2.5 m above the soil surface. A short and intense downpour, with an intensity of 0.8 or 2.0 mm/min and the duration of 60 or 30 min, was applied. The landscape was modeled using matrix loess collected from the Shunyi District, Beijing. The 50% diameter of soil particles, d_{50}, was 52.2 μm, and the specific gravity, γ_s, was 2.56. An experimental model landscape, with a steep slope of 60°–80° and the gentle slope of 3°, was developed. Soil slope was prepared by hand patting to generate a 'smooth' roughness to ensure a regular and original micro relief. For an experimental group with the same initial landform, 5–10 events of rainfalls were applied to the slope in turn. An equal period, 12 h or so, was kept after each rainfall to ensure the approximate value of initial water content. The topography meter was used to monitor the mass movement of the failure surface under rainfall simulation.

4.3   Data Processing

4.3.1   Method to Calculate Gravity Erosion and Hydraulic Erosion

Based on the contour map obtained from the topography meter, a 3D stereogram was established with ArcGIS, and then the slope parameters such as volume, projection area and point coordinates were obtained. The soil loss was calculated using the following equations:

$$g_{(i,j)} = v_{1(i,j)} - v_{2(i,j)} \tag{4.1}$$

$$T_j = V_{j1} - V_{j2} \tag{4.2}$$

$$H_j = T_j - \sum_{i=1}^{N} g_{(i,j)} \tag{4.3}$$

where i represents the sequence number of the failure incident during a rainfall; j represents the sequence number of the rainfall for a certain initial landform; $g_{(i,j)}$ is the volume of an individual failure mass; $v_{1(i,j)}$ and $v_{2(i,j)}$ are the slope volumes in the scope of the incident in the moments before and after the failure, respectively; T_j is the total amount of soil loss during a rainfall event; V_{j1} and V_{j2} are the slope volumes in the moments before and after the rainfall, respectively; H_j is the total amount of soil loss caused by water flowing during the rainfall event; and N is the number of failure events during a rainfall.

4.3.2   Method to Measure Slope Volume

The slope volume is measured with the following technical route as shown in Fig. 4.4. Involved software includes the Chinese versions of QQ Video 2013 v4, Photoshop 6.0, R2V 5.5 and ArcGIS 10.0.

1. Getting screenshots in QQ Video and marking landslide traces in Photoshop

Case No. 1: Dealing with an individual collapse. Use the software QQ Video to open the test video, and tap the keyboard shortcuts Alt+A to get the screenshots

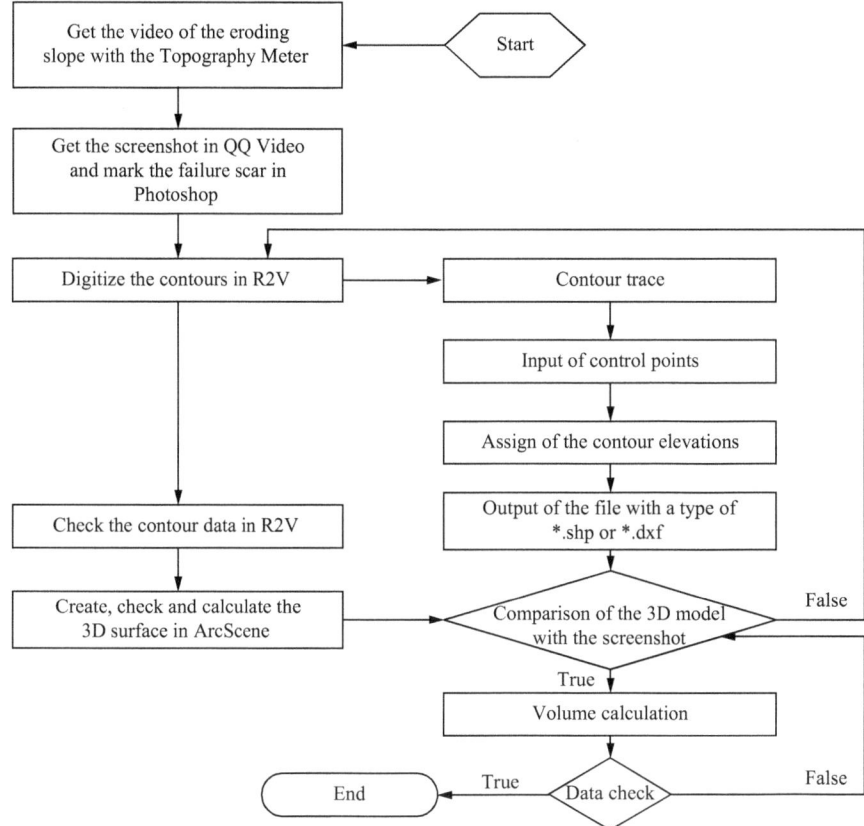

Fig. 4.4 Outline of methodological procedure for calculating the slope volume. The amount of soil loss is the slope volume within the frame at the beginning of the event subducing that at the end of the event. However, for the individual mass failure, boundaries only enclose the range of the individual landslide scar being investigated, but for the total soil loss, the virtual boundaries should include the scope of all of the soil erosion during the rainfall

seconds before the failure starts and soon after the incident finishes. Then add the virtual boundaries to the screenshots in Photoshop, i.e., the white frames shown in Fig. 4.5a, b. The boundaries should enclose the traces caused by the individual collapse. Surely, the frames in the two images should be in the same size and position.

Case No. 2: Dealing with a whole erosion under rainfall. Use the software QQ Video to get the screenshot before the start of sprinkling as above. Then go to the scene to get the screenshot after the rainfall. Open the screenshots in Photoshop, and add the virtual boundaries to the screenshots including the scope of all soil erosion during the rainfall, shown by the white frames in Fig. 4.6a, b.

2.   Digitizing the topographic map in R2V

Load the picture with virtual boundary in R2V, and then portray the contours, set the contour heights and input the control points. Finally, output the above file with the

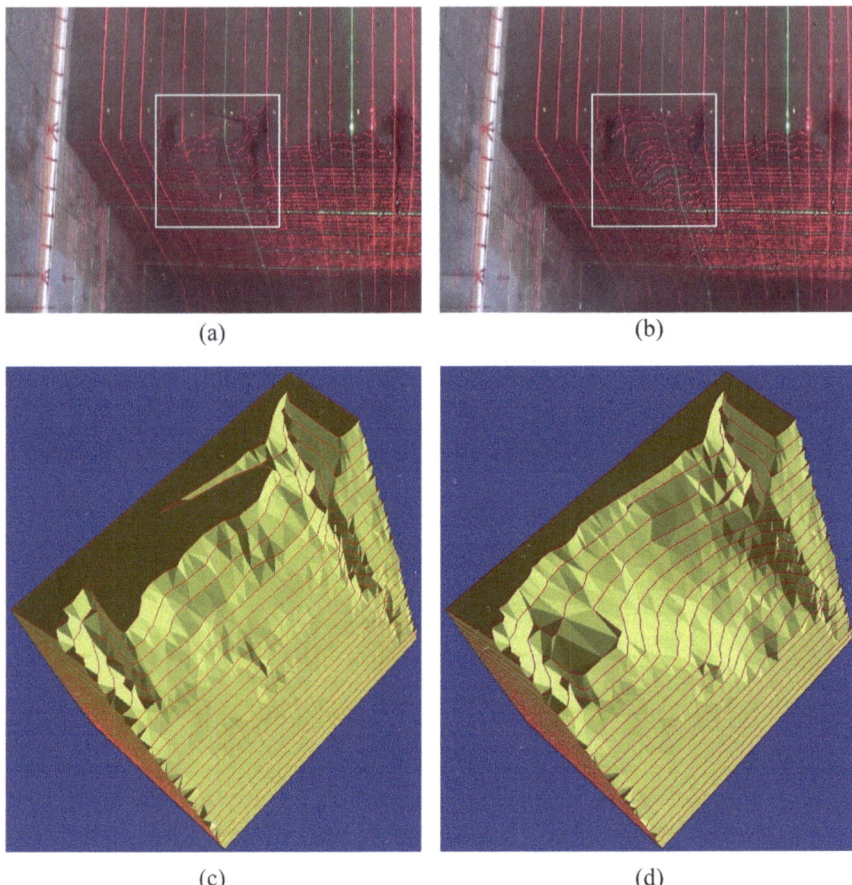

Fig. 4.5 Comparison of the model slope and the three-dimensional vector for an individual collapse. **a** The input image, seconds before the failure starts. **b** The input image, seconds after failure finishes. **c, d** are the resulting 3D surface models within the white rectangles corresponding to **a, b**, respectively. The volume difference of the two slopes shown in **c, d** is the volume of the individual mass failure

format of *.shp or *.dxf. Since the measurement accuracy of failure volume is closely correlated to the contour precision, the tracing contours must completely match the laser lines in the screenshot.

3. Creating a TIN surface and calculating the volume with ArcGIS

Open ArcScene of the ArcGIS, import the vector file *.shp or *.dxf, and form the TIN surface of the slope. Check the surface by comparing with the screenshot, and then calculate the volume of a given surface with the command Surface Volume. Otherwise, examine and revise the file *.prj in R2V once again.

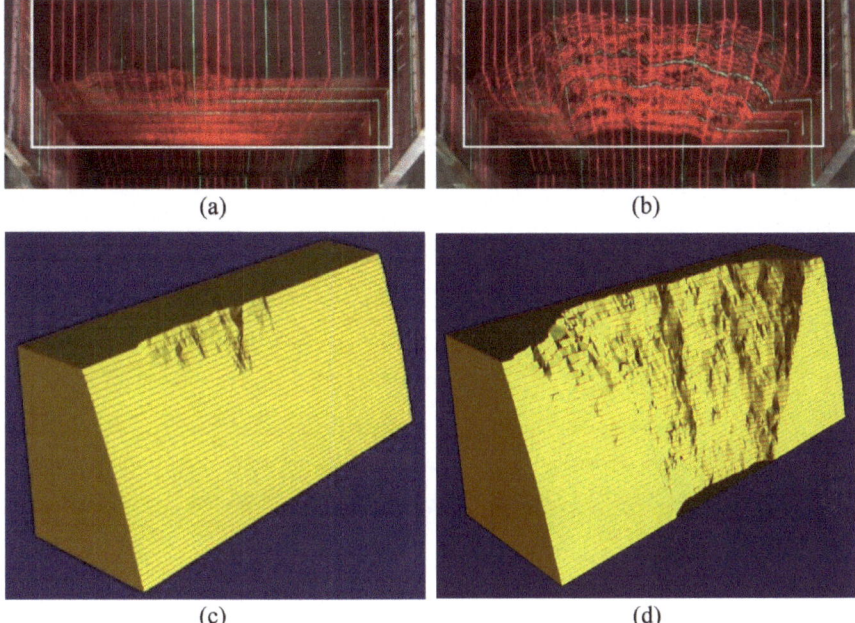

Fig. 4.6 Comparison of the model slope and the three-dimensional vector for a rainfall event. **a** The input image before the start of sprinkling. **b** The input image 60 min later when the rainfall was over. **c** and **d** are the resulting 3D surface models corresponding to **a**, **b**, respectively. The volume difference of the two slopes shown in **c**, **d** is the total amount of soil loss during the rainfall event

4.4 Results and Discussion

4.4.1 Calibration Test for Measurement of the Slope Volume and Coordinates

For the laboratory calibration tests on the wooden brae, the relative errors among the volumes observed by the topography meter and those of the conventional instruments were all within 10% for the five landform models. In the test, the volume of the landform was about 24,000 cm^3 and the slope gradients were from 35° to 75°, where the maximum error was 9.0%, and the minimum was 0.4%. For the field calibration tests on the natural hillside with the range of 120 cm × 140 cm, the mean square deviations of the ten selected points on x, y and z axis were 0.99 cm, 0.71 cm and 0.18 cm, respectively. Figure 4.7 and Table 4.1 show that the slope volumes and point coordinates measured with the topography meter were very close to those measured with the conventional instruments of steel rule and level instrument. These experiments demonstrate that the system accuracy remains acceptable and the topography meter is applicable to study steep slopes. The measurement differences

Fig. 4.7 Volumes of the brae in various slope gradients observed with the topography meter and the conventional instruments, respectively. Here, the conventional instruments mean the steel rule, the level instrument, etc

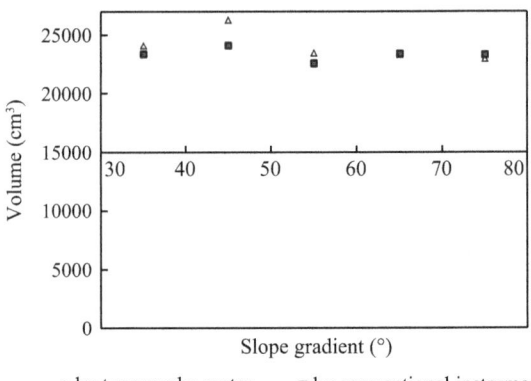

△ by topography meter   ■ by conventional instruments

Table 4.1 Data of check points observed by a level and the topography meter, respectively

Check points	Observed with a level/cm	Observed with the topography meter/cm	Check points	Observed with a level/cm	Observed with the topography meter/cm
1	(20.00, 80.00, 36.52)	(20.94, 80.90, 36.13)	6	(70.00, 40.00, 20.42)	(69.35, 41.81, 20.86)
2	(20.00, 100.00, 44.92)	(20.77, 100.38, 45.14)	7	(70.00, 50.00, 24.52)	(68.53, 50.97, 24.00)
3	(30.00, 60.00, 21.72)	(31.36, 60.39, 21.05)	8	(70.00, 80.00, 51.22)	(69.12, 80.24, 51.12)
4	(40.00, 130.00, 56.97)	(40.58, 128.83, 57.07)	9	(70.00, 100.00, 65.60)	(68.56, 100.61, 65.98)
5	(50.00, 90.00, 39.52)	(49.32, 90.13, 39.33)	10	(80.00, 90.00, 68.32)	(79.49, 89.97, 69.00)

were mainly attributed to the exactness of the shooting angle of the camera and the vector conversion of images.

4.4.2   Case Study for an Individual Failure

Over 120 rainfall simulation events have been completed with the topography meter, which again confirmed the feasibility and reliability of this technique. For the slopes with the same height, the amount of collapse increased with the enlargement of the slope gradient, but the amounts of landslides and the total gravity erosions were different. We have used the quantitative data to confirm that for the slopes with the same slope gradient, the amount of collapse increased with the enlargement of the slope height, but the amounts of landslides were different (Xu and Zhao 2014). We also counted the ratio of the gravity erosion to the total slope erosion, and affirmed that gravity erosion is more dangerous than hydraulic erosion on the steep loess slope, as briefly shown in the following text. All results seem rational and reliable.

Table 4.2 illustrates a sample of measured failure bulks in 6 rainfall events. Any collapse with volume more than 100 cm^3 was accurately measured. According to the forms of mass failure and the failure surfaces, erosion type could be identified, and then detailed statistics of gravity erosion, including planar block slide, rotational slump, and earth flow, were figured out. In Fig. 4.5, a typical scenario and the resulting 3D digital model of the failure surface was illustrated. High resolution is achieved through labeling the grid surface, and the results obtained by the topography meter are quite consistent with the real phenomena *in situ*. Different from any other monitoring device, the topography meter only deals with the screenshots of the concerned erosion incidents although it records the whole erosion process during the rainfall. Thus the

Table 4.2 Amounts of the individual mass failures: a sample

Experimental group	Rainfall event	Volume (cm^3)[a]	Run designation					Events of failure behaviors
			1	2	3	4	5	
L60-1.5-80d	120924-1	v_1	7253.7	12,252.1	4295.6	–	–	3
		v_2	6902.8	11,931.7	4116.0	–	–	
		g	351.0	320.5	179.6	–	–	
	120924-2	v_1	11,639.6	19,261.0	14,537.5	25,841.3	…[b]	21
		v_2	10,688.0	18,334.5	14,312.2	25,323.4	…	
		g	951.7	926.6	225.3	517.9	…	
	120925-3	v_1	15,131.0	23,764.3	–	–	–	2
		v_2	13,991.0	18,695.9	–	–	–	
		g	1140.0	5068.4	–	–	–	
	120925-4	v_1	33,0097.6	112,638.1	633,296.2	836,295.3	…	7
		v_2	289,535.0	104,731.1	496,692.4	680,258.9	…	
		g	40,562.6	7907.0	136,603.8	156,036.4	…	
	120926-5	v_1	–	–	–	–	–	0
		v_2	–	–	–	–	–	
		g	–		–	–	–	
	120926-6	v_1	6874.7	84,446.9	13,596.1	134,468.7	…	8
		v_2	5783.5	63,571.9	11,323.2	128,463.3	…	
		g	1091.2	20,875.0	2272.9	6005.4	…	

The data were obtained from the second rainfall of the L60-1.5-80d experimental group, of which the rainfall intensity was 0.8 mm/min and the duration was 60 min, and the slope height and gradient of the initial landform before the first rainfall were 1.5 m and 80°, respectively. The initial terrain before the second rainfall event was that 12 h after completion of the first rainfall

Notes a. v_1 and v_2 denote the slope volumes in the moments before and after the mass failure, respectively. g is the volume of the failure mass, and $g = v_1 - v_2$

b. Only are the data of four events of mass failures listed as samples in the table although more than five events happened in the rainfalls 120924-2, 120925-4, and 120926-6

workload is relatively minor, and the method is faster and more precise. Hence, the volume of "pure" gravity erosion could be also calculated by clearly differentiating gravity erosion from hydraulic erosion.

4.4.3 Case Study for the Total Amount of Erosion

Figure 4.6a, b show the screenshots in the experiment group L60-1.5-80d, of which the initial slope height was 1.5 m and the initial gradient was 80°, at the beginning of the 2nd rainfall and soon after a 60 min rainfall, respectively. Figures 4.6c and d are the three-dimensional models formed according to the screenshots. According to the three-dimensional vectors as mentioned above, the total amount of soil erosion during the rainfall was figured out, having the number of 239.16×10^3 cm^3.

For experiment group L60-1.5-80d, the calculated results of gravity erosions are illustrated in Table 4.3. For instance, in the 3rd rainfall event, gravity erosion with volumes more than 100 cm^3 occurred 2 times, and the total amount reached 6.21×10^3 cm^3, all of which were completed in a very short time. Simultaneously, the amount of hydraulic erosion was 4.64×10^3 cm^3, which lasted for the whole period of rainfall simulation. In the 2nd, 3rd and 4th rainfall events, the amount of gravity erosion accounted for up to 91.89, 57.21 and 66.47% of the total erosions, respectively. As a result, gravity erosion is more dangerous than hydraulic erosion on the steep loess slope.

4.4.4 Limitations and Future Developments

Just like any other structured-light 3D surface measuring techniques, the topography meter casts straight lights on the slope and has difficulty in handling the cavities

Table 4.3 Total amount of soil loss during a rainfall: a sample

Experimental group	Rainfall event	Events of gravity erosion	Soil loss during rainfall (10^3 cm^3)			$\frac{G_1}{T}$ (%)
			Total erosion (T)	Gravity erosion (G_1)	Hydraulic erosion (H)	
L60-1.5-80d	120924-1	3	13.09	0.85	12.23	6.50
	120924-2	21	260.27	239.16	21.11	91.89
	120925-3	2	10.85	6.21	4.64	57.21
	120925-4	7	538.04	357.63	180.41	66.47
	120926-5	0	22.36	0.00	22.36	0.00
	120926-6	8	240.61	105.75	134.86	43.95

The data were obtained from the second rainfall of the L60-1.5-80d experimental group

where the laser could not reach. Though we are fortunate that we didn't meet such a perplexity in our studies, it's still a limitation of the technique. With the improvement of the construction of the laser source and the enlargement of the monitoring range, the apparatus could be updated and this disadvantage could be eliminated. Another serious shortcoming of the topography meter comes from the laboratory image processing. The equipment of a camera similar to the traffic light, which could autonomously take a snapshot as the failure occurs, could probably solve this problem. At that time, the volume of individual collapse could be directly shown on the screen, as the laser stripes are auto-vectored with ArcGIS, R2V, or other professional software (e.g., Fechteler and Eisert 2009).

The topography meter could be applied in the field test as an appropriate observation environment, e.g., a portable tent, is provided. Presently the authors are conducting a site experiment using the topography meter on the Loess Plateau of China. In the near future, our findings could be generalized to different regions which might have different hydrological responses (e.g., Zhang et al. 2014).

4.5 Conclusions

A novel structured-light 3D surface measuring technique was developed to quantitatively measure time-variable gravity erosion on the steep loess slopes. The topography meter, mainly consisted of a camera, laser source, and positioning device, has a function of recording slope deforming process under laser marking. Hence, 3D vector images at the moment of erosion incident could be obtained. Erosion data, including the amount of each failure mass, the total amount of soil loss eroded by overland flow, etc., could accordingly be calculated. A series of calibration tests indicated that the apparatus was at a sufficiently high accuracy, enough for the mechanism exploration of slope erosion, especially for that caused by gravity. More than 120 rainfall simulation events were also measured, further testifying its feasibility and reliability. The apparatus is very promising in field observations after improving the laser source and the imaging process.

References

Abellán A, Vilaplana J M, Martínez J. 2006. Application of a long-range Terrestrial Laser Scanner to a detailed rockfall study at Vall de Núria (Eastern Pyrenees, Spain). Engineering Geology, 88(3): 136–148.

Casula G, Mora P, Bianchi M G. 2010. Detection of terrain morphologic features using GPS, TLS, and land surveys: "Tana della Volpe" blind valley case study. Journal of Surveying Engineering, 136(3): 132–138.

Du J C, Teng H C. 2007. 3D laser scanning and GPS technology for landslide earthwork volume estimation. Automation in Construction, 16(5): 657–663.

Du Z T, Lu L, Zhang W H, et al. 2013. Measurement of the velocity inside an all-fiber DBR laser by self-mixing technique. Applied Physics B, 113(1): 153–158.

Fechteler P, Eisert P. 2009. Adaptive colour classification for structured light systems. IET Computer Vision, 3(2): 49–59.

Geng J. 2011. Structured-light 3D surface imaging: a tutorial. Advances in Optics and Photonics, 3(2): 128–160.

Gokceoglu C, Sezer E. 2009. A statistical assessment on international landslide literature (1945–2008). Landslides, 6(4): 345–351.

Hancock G R, Crawter D, Fityus S G, et al. 2008. The measurement and modelling of rill erosion at angle of repose slopes in mine spoil. Earth Surface Processes and Landforms, 33(7): 1006–1020.

Hergarten S. 2012. Topography-based modeling of large rockfalls and application to hazard assessment. Geophysical Research Letters, 39(13): L13402.

Meng C, Janssen M H. 2015. Measurement of the density profile of pure and seeded molecular beams by femtosecond ion imaging. Review of Scientific Instruments, 86(2): 023110.

Ohnishi Y, Nishiyama S, Yano T, et al. 2006. A study of the application of digital photogrammetry to slope monitoring systems. International Journal of Rock Mechanics and Mining Sciences, 43(5): 756–766.

Peng T. 2007. Algorithms and models for 3-D shape measurement using digital fringe projections. Maryland: University of Maryland.

Rieke-Zapp D H, Nearing M A. 2005. Digital close range photogrammetry for measurement of soil erosion. Photogrammetric Record, 20(109): 69–87.

Scherer M, Lerma J L. 2009. From the conventional total station to the prospective image assisted photogrammetric scanning total station: comprehensive review. Journal of Surveying Engineering, 135(4): 173–178.

Thomas T, Rossman G R, Sandstrom M. 2014. Device and method of optically orienting biaxial crystals for sample preparation. Review of Scientific Instruments, 85(9): 093105.

Vondrak J, Machotka R, Podstavek J, et al. 2006. Use of fresnel enantiomorphic diffraction in surveying laser methods. Journal of Surveying Engineering, 132(2): 71–76.

Wang Y C, Liu K, Lau D L, et al. 2010. Maximum SNR pattern strategy for phase shifting methods in structured light illumination. Journal of the Optical Society of America A, 27(9): 1962–1971.

Wolf P R. 2002. Surveying and mapping: History, current status, and future projections. Journal of Surveying Engineering, 128(3): 79–107.

Xu X Z, Zhao C. 2014. A laboratory study for gravity erosion of the steep loess slopes under intense rainfall. Proceeding of 11th International Conference on Hydroscience and Engineering, Hamburg: 709–715.

Xu X Z, Zhang H W, Wang W L, et al. 2015. Quantitative monitoring of gravity erosion using a novel 3D surface measuring technique: validation and case study. Natural Hazards, 75(2): 1927–1939.

Zhang S. 2005. High-resolution, real-time 3-D shape measurement. New York: Stony Brook University.

Zhang C, Wang D G, Wang G L, et al. 2014. Regional differences in hydrological response to canopy interception schemes in a land surface model. Hydrological Processes, 28(4): 2499–2508.

Chapter 5
How to Conduct an Experiment in the Field: A Portable Laboratory

Abstract Observation of the gravity erosion in the field with strong sunshine and wind poses a challenge. Here, a novel topography meter together with a movable tent addresses the challenge. With the topography meter, a 3D geometric shape of the target surface can be digitally reconstructed. Two methods can be used to obtain a relatively clear video, despite the extreme steepness of the slopes. One method is to rotate the laser source away from the slope to ensure that the camera sightline remains perpendicular to the laser plane. Another way is to move the camera farther away from the slope in which the measured volume of the slope needs to be corrected; this method will reduce the distortion of the image. In addition, the installation of tent poles with concrete columns helps to surmount the altitude difference on steep slopes. Results observed by the topography meter in real landslide experiments are rational and reliable.

Keywords A portable laboratory · Field landslide experiments · Topography meter · Slope erosion

5.1 Difficulties to Conduct a Rainfall-Simulation Experiment *in Situ*

Applicable and precise instruments are vital to obtain credible data in experiments. Recently, a structured-light 3D surface-measuring instrument, the MX-2010-G topography meter, was designed and manufactured to observe the slope behavior under rainfall simulation. Evaluation tests showed that the errors among the volumes observed by the Topography meter and those of the conventional instruments were within 10% (Xu et al. 2015). However, the above-mentioned paper was focused on data processing and calibration testing. It did not give detailed information on the structure of the 3D surface measuring instrument. In addition, the definition of laser lines recorded by the camera had a high requirement for the ambient light. If the ambient lighting was sufficiently strong, the contrast of the laser footprints would decrease on the sloping terrain, and the image would become vague. Whether it is possible to use the topography meter for conducting site-specific surveys of landslides has become an urgent question to be resolved.

© Science Press and Springer Nature Singapore Pte Ltd. 2020
X. Xu et al., *Experimental Erosion*,
https://doi.org/10.1007/978-981-15-3801-8_5

Gravity erosion, the mass failure on steep slopes that is triggered by self-weight, contrasted with other types of soil erosion requiring physical impetus of wind or water, is among the most important natural hazards in mountainous regions (Hergarten 2012; Gokceoglu and Sezer 2009). Gravity erosion tends to happen as an episodic event in which large sections of the steep slope fail. However, to monitor the time-variable process of an individual mass failure is very difficult because of the randomness and suddenness of such an event. It is rather difficult to conduct a field survey of soil and water conservation on the Loess Plateau of China. During the natural rainfall events, the site-specific observation of gravity erosion is almost impossible, because we are unable to predict when and where the gravity erosion will occur. Hence, what we could do is to simulate the landslide processes by employing a conceptual slope under rainfall simulation. However, the representative catchments on the Chinese Loess Plateau for scientific research are generally far from urban areas, lack water and electricity, and have deep ditches, steep slopes, inconvenient transportation, and high-strength winds. The special situation poses another severe challenge to the site survey. Here, we designed a removable house that could be assembled in the field to conduct site-specific tests based on the terrain of the Loess Plateau. The room provided the same conditions for simulations and observations as those in the laboratory. We also presented a detailed illustration of the MX-2010-G topography meter to dynamically monitor the slope behavior under rainfall simulation. The design optimization and fabrication methods of the topography meter and the laboratory would be expected to be useful for congeneric device applications.

A MX-2010-G topography meter has been used in real gravity erosion experiments. Rainfall simulation tests with the unsheltered initial landforms and rainfalls were carried out on the actual site of the Liudaogou Catchment, Loess Plateau, as shown in Fig. 5.1a, b.

(a)   (b)

Fig. 5.1 A MX-2010-G topography meter was used in the field study of the gravity erosion on the Loess Plateau. **a** A movable lab for field study of gravity erosion on the Loess Plateau of China; **b** The topography meter used in the field tests

5.2 Design of the Measurement System

5.2.1 A Movable Tent for Field Study

The experimental house built in the field was adapted from a large bivouac tent and could be assembled. Generally, the tourism tents available on the market could not be used directly on steep slopes. Furthermore, the tents could not resist strong winds due to their slender steel poles. In this study, besides the iron wires provided by the manufacturer of the tent, many mooring ropes were used to fix the tent, together with a high-strength concrete pole based on a large pier on the downhill side, as shown in Fig. 5.2a, b. All of the measures made the tent very steady.

In addition, four sturdy columns were installed in the lower reaches of the slope. The tent poles at the lower-lying area were connected with the columns that had several equidistant connecting holes along their height directions. When the camp pole was fixed with the bolts in different connecting holes on the column, various lengthening effects could be attained. The scheme to lengthen the tent poles with the concrete columns completely solved the problem of the altitude difference on the steep slope (Fig. 5.2a). If the site terrain was prepared, the tent could be built with a surprising speed in 2–3 days.

5.2.2 Operating Principle of the MX-2010-G Topography Meter

A structured-light 3D surface-measuring instrument, the MX-2010-G topography meter, was designed and manufactured to observe the slope behavior under rainfall simulation. In our experimental system, the topography meter's setup consists of the following conventional components: sight calibrator, laser source, camera, and positioning device. Figure 5.3 shows a representative scheme for quantitative monitoring of gravity erosion with the MX-2010-G topography meter. The horizontal stripe pattern with a 3.0 cm contour interval was generated by a laser source, and the pattern was recorded by the camera with a sighting direction perpendicular to the laser planes. A camera with a collimator was mounted on the bracket. The collimator could submit a line-shape beam parallel to the camera sightline. As the slope deformed over time, the process would be recorded on video and then imported into the computer to acquire a snapshot image at a particular time. Given an elevation for every contour in ArcGIS, the 3D geometric shape of the target surface was digitally reconstructed and then the slope parameters, including the volume, projected area, and gradient distribution, could be determined. Thus, we could obtain the volume of gravity erosion and many other erosion data by comparing the slope geometries in the moments before and after the erosion incident. In this study, the topography meter covered a monitoring range of 3.0 m × 2.0 m.

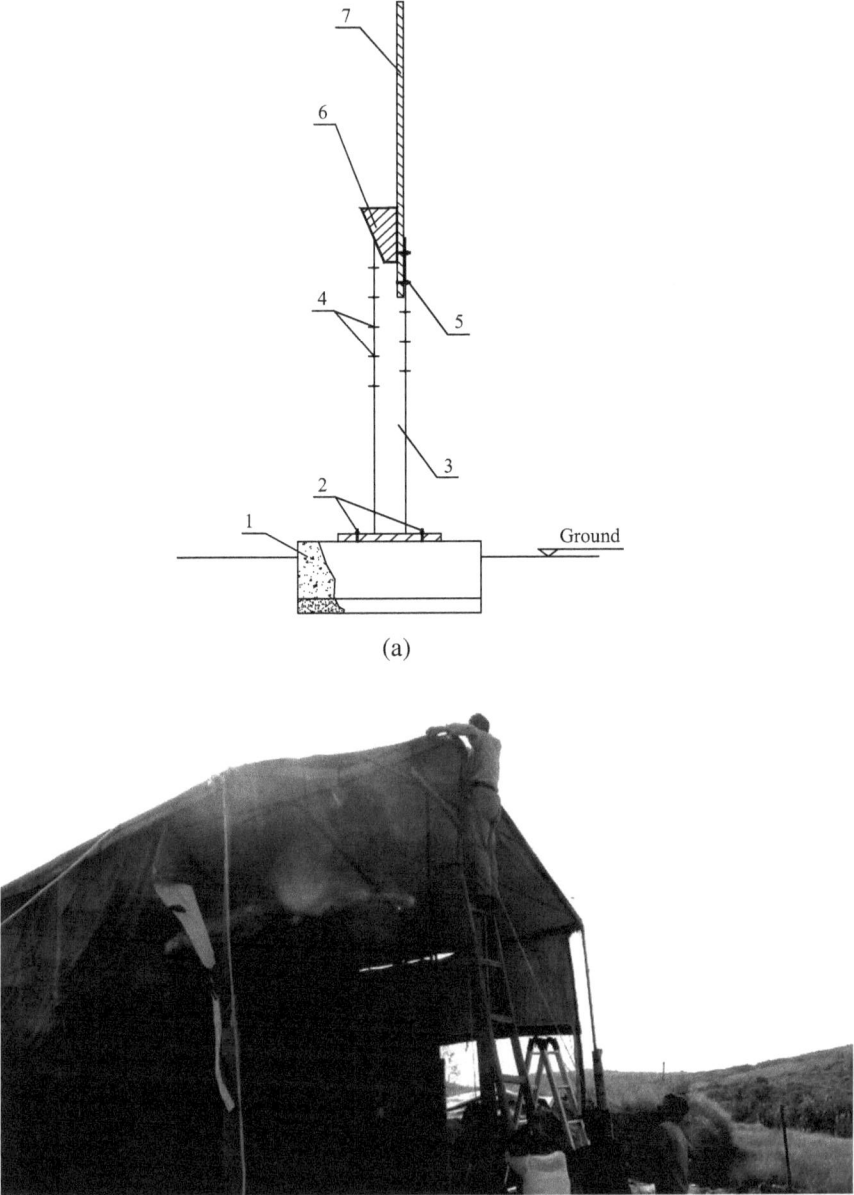

(a)

(b)

Fig. 5.2 A tent-style removable house designed for the field study of the gravity erosion on the Loess Plateau. **a** A lengthened column; **b** The construction site. Keys: 1. Base; 2. Positioning bolt; 3. Concrete column; 4. Connecting holes; 5. Set bolt; 6. Wooden wedge; 7. Tent pole

Fig. 5.3 A MX-2010-G topography meter used in a concrete landslide experiment. Keys: 1. Laser source; 2. Camera with a collimator; 3. Positioning device with a spot laser; 4. Bracket; 5. Conceptual model slope with navigation dot markers and equidistant projected fringes

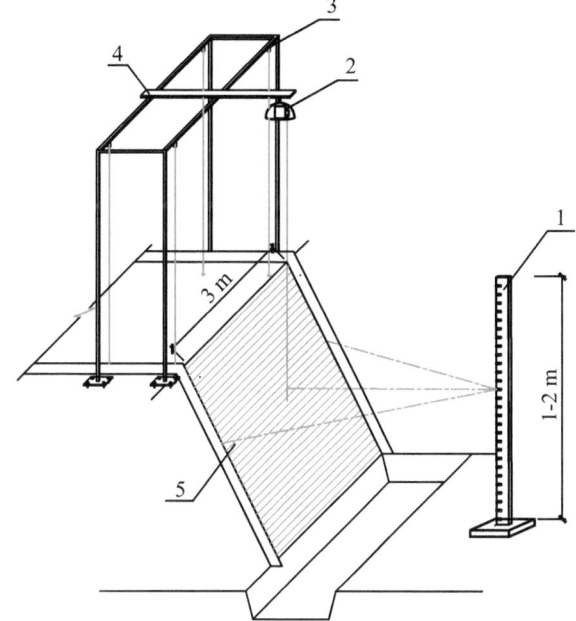

Before the commencement of any slope behavior survey, we should adjust the camera and make its sightline perpendicular to the laser planes using the sight calibrator. The calibrator was a ⌒\-shaped open container with three opaque walls, as shown in Fig. 5.4. A group of parallel black lines spacing apart by a distance of 3 cm were drawn on the transparent wall of the box. A reflecting mirror was set at the bottom of the prismatic box. The operator held the sight calibrator under the camera of the topography meter as the calibration experiment began. Then the parallel laser planes were shot to the open mouth and projected onto the three walls. Meanwhile, the laser lines from the collimator were thrown on the mirror at the bottom of the sight calibrator. When the laser plane was perpendicular to the camera sightline, the footprints of the laser plane were overlapped with black lines of the barrel wall, and the spot lines from the collimator were directly reflected back to the collimator. The calibrator could accurately examine the sight direction. Nevertheless, manipulating the calibrator could be difficult because the operator had to hold it. If a trestle was designed for the collimator in the future, the operation would become more convenient.

5.2.3 Structure of the MX-2010-G Topography Meter

The laser source was mounted on a rugged pedestal (Fig. 5.5), including the line laser modules, module fasteners, a girder, and a cover. The line laser modules, also

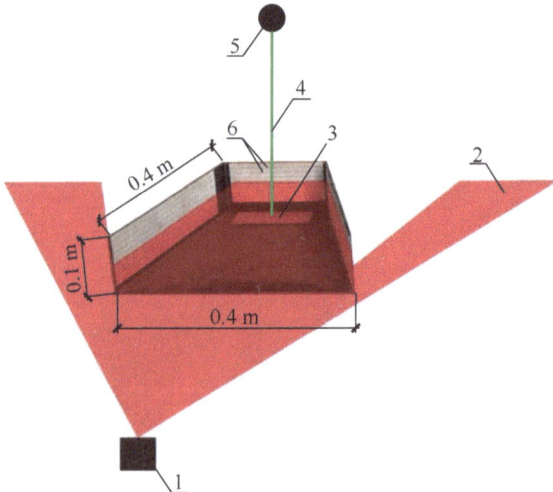

Fig. 5.4 A scheme to calibrate the angle between the laser plane and the camera sightline. Keys: 1. Laser source; 2. Equidistant laser planes; 3. Mirror; 4. Spot positioning laser; 5. Camera with a collimator; 6. Equidistant lines on the box wall

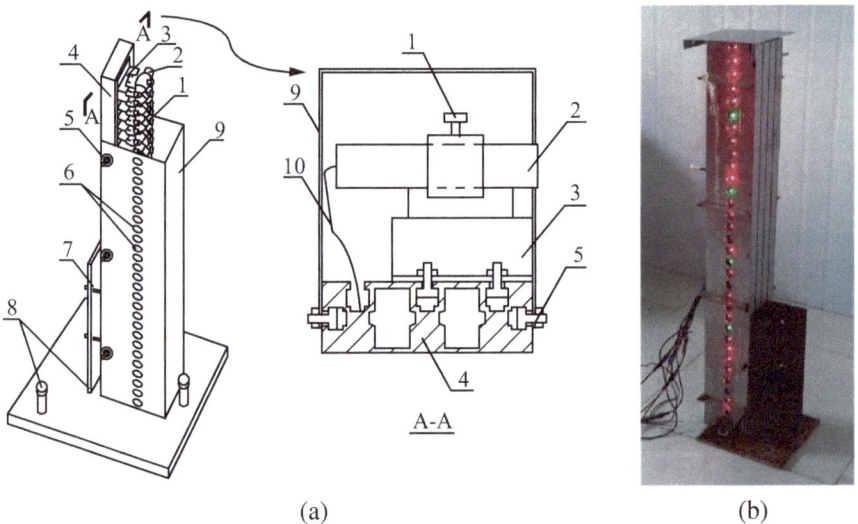

(a)   (b)

Fig. 5.5 Detailed structure of the laser source. **a** Schematic view; **b** Product photo. Keys: 1. Adjustable bolt; 2. Line laser module; 3. Module fastener; 4. Girder; 5. Fixing bolt; 6. Embrasure; 7. T-styled bolt; 8. Pedestal; 9. Cover; 10. Electric wire

referred to as line laser heads, were self-contained green, infrared, and red line laser generators with an integrated laser driver circuit, optics, and laser diode. They were suitable for different applications such as alignment, positioning, and measurement, and were widely used in education, military, manufacturing, and medical treatment, etc. To expediently compute the laser number on the laser source, five modules that each emitted a red laser were set at intervals for each module that emitted a green laser. The line laser modules were powered by the DC electrical source. The laser line generator used in the MX-2010-G topography meter offered a compact, durable product that was designed and built to perform reliably even under adverse industrial conditions. The diameter of the laser head was only 1.6 cm, and the expected lifetime was up to 10,000 h.

When the topography meter was carried to the field to conduct the site survey, the position of the line laser module in the laser source might be changed due to the jolts and vibrations in transit. As a result, the laser planes would become nonparallel or the intervals of them would become unequal. Hence, we should adjust the positions of the modules to make the lasers parallel and equidistant before commencement of the slope behavior survey if the topography meter had been re-installed in the test site. As shown in Fig. 5.5, the fixed bolt, which was on the module fastener, was connected with the mainboard. To adjust the relative distance of the laser planes, the fixed bolt should be modulated to make the fastener move back and forth on the mainboard. The line laser module fixed in the ring of the module fastener could emit a group of parallel laser lines. If the line laser module moved out of position, modify the adjustable bolt to swing or rotate the line laser module slightly, and then make the laser plane return to the parallel state.

5.3 Applications in the Landslide Experiments

5.3.1 Results of the Landslide Experiments

The landscape simulator consisted of a rainfall simulator and a slope model covering an area of 3.0 m × 3.0 m. An experimental model landscape, with a steep slope of 60°–80° and the gentle slope of 3°, was made of loess by hand patting. The conceptual landscape in the field was "cut" without disturbing the slope underground. In other words, the initial landform in the field study kept the original texture and density, although the surface was cut to be smooth. Underlying the surface of the tested slope was a kind of sandy Holocene loess. The dry bulk density ρ_d was 1.54 g/cm^3 and the median size d_{50} was 143.0 μm. A short and intense downpour, with an intensity of 0.8 mm/min and a duration of 60 min, was applied. In turn, 5–10 events of rainfalls were applied to the slope. An equal period, 12 h or so, was kept after each rainfall to ensure the approximate value of initial water content. The MX-2010-G topography meter was used to monitor the mass movement of the failure surface under rainfall simulation (Fig. 5.1b). Furthermore, a large tent was set up in the experimental spot to keep out sunshine and winds. At the lower right corner of Fig. 5.1b, an MX-2010-G topography meter was working.

The 3D digital models of the unsheltered surface of the ground were valuable supports for geomorphological studies because they provided scientists with a full and immediate perception of the represented ground surface trends (Zhang et al. 2014, 2015; Casula et al. 2010). Figure 5.6a, b illustrate a typical scenario and the resulting 3D digital model of the failure surface, respectively. High resolution is achieved through labeling the grid surface, and the results obtained by the topography meter are quite consistent with the real phenomena. Different from any other monitoring devices, the topography meter only deals with the screenshots of the concerned erosion incidents, although it records the entire erosion process during the rainfall. Thus, the workload is relatively minor and the method is quicker and more precise.

The ratio (R) of the amount of gravity erosion to that of the total erosion on the Loess Plateau, China was considerable. Experimental data in Table 5.1 illustrate that the amounts of the gravity erosion during rainfall accounted for 67% of the amounts of total erosion. Nevertheless, the hydraulic erosion lasted for the entire period of rainfall simulation, but the gravity erosion happened in a very short time. Thus, gravity erosion is more dangerous than hydraulic erosion on the steep loess

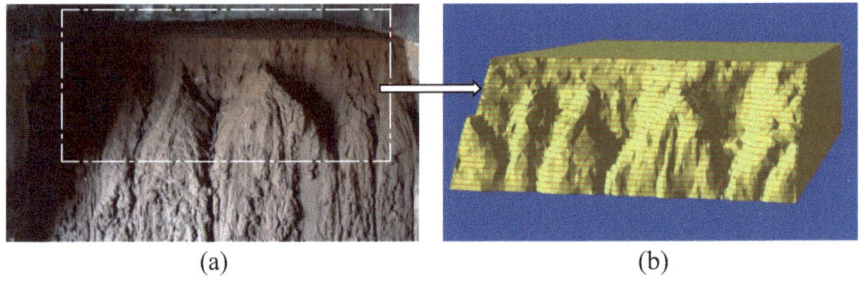

(a)   (b)

Fig. 5.6 Comparison of the real slope and the three-dimensional vector for an individual collapse. **a** Photo of the experimental scene; **b** 3D surface model of the eroding slope

Table 5.1 Soil loss in the experiment L60-1-70d, of which the initial slope height was 1 m and initial gradient was 70°

Rainfall event		Total erosion, T (10^3 cm^3/m)	Gravity erosion, G (10^3 cm^3/m)	Hydraulic erosion, H (10^3 cm^3/m)	$R = G/T$ (%)
Field experiment	130826-1	0.86	0.69	0.17	80
	130826-2	21.59	16.10	5.49	75
	130826-3	29.83	21.95	7.87	74
	130827-4	6.17	3.23	2.95	52
	130828-5	3.49	1.91	1.58	55
	Average	12.39	8.77	3.61	67

slope. In addition, according to the investigation by the Management Board for the Middle Reaches of the Yellow River (MBMRYR 1993), the ratios R values in the Nanxiaohegou Catchment, Lv'ergou Catchment, and Jiuyuangou Catchment were 58, 68 and 20%, respectively. Results in this study are similar to the research conclusions mentioned above.

5.3.2 Methods to Obtain Relatively Clear Video

Typically, the topography meter could measure the gravity erosion on the slope with a gradient of 30°–70°. In the present laboratory condition, the laser projection received in the camera is very clear. However, on the Loess Plateau of China, some gully banks are extremely steep, sometimes more than 80°. One method to obtain a relatively clear video is to rotate the laser source away from the slope, based on the premise of ensuring the camera sightline remains perpendicular to the laser plane, as shown in Fig. 5.7. As a result, the slope with projected lasers shown in the camera becomes relatively gentle so that the best observation effect can be obtained. In Fig. 5.7, the

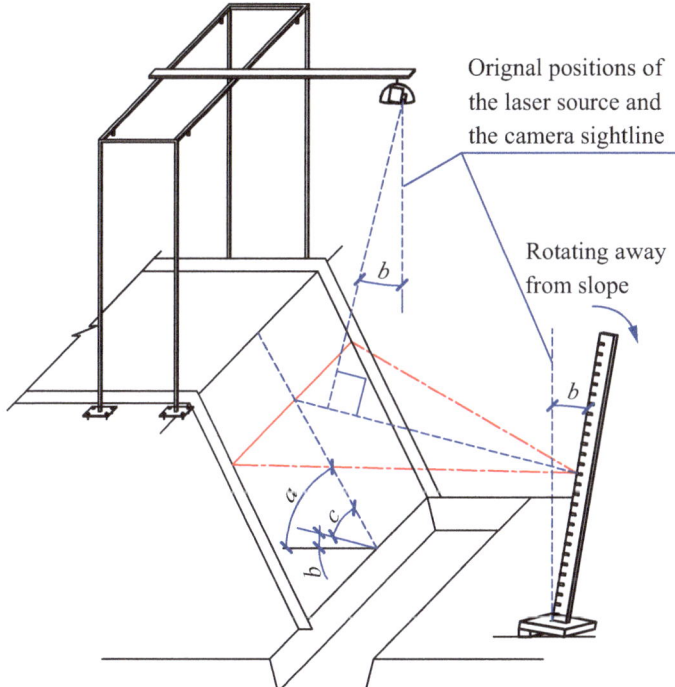

Orignal positions of the laser source and the camera sightline

Rotating away from slope

Fig. 5.7 A method to obtain a clear and undistorted video for the very steep slope. The laser source is rotated away from the slope based on the premise of ensuring that the camera sightline remains perpendicular to the laser plane

slope angle a is equal to the angle between the slope and the laser plane when the laser source is perpendicular to the ground. However, if the laser source were rotated away from the slope with an angle of b, the angle between the slope and the laser plane, c, would be equal to $(a - b)$. In this case, the observed slope does not deform because the camera sightline is kept vertical to the laser plane; that is, no discrepancy occurs between the real volume and the measured volume of the slope.

Another way to obtain a relatively clear video is to move the camera away from the slope. Nevertheless, in the condition, the camera sightline is not perpendicular to the laser plane, and the measured volume should be corrected. As shown in Fig. 5.8, a slant camera—i.e., the angle between the camera sightline and the laser plane, d, was less than 90°—could also show relatively clear video. In this way, the image of the steep slope in the camera became relatively gentle so that the laser projection could be easily read. Nevertheless, the slope volume obtained with the slanted camera must be corrected to minimize the image distortion. Thus, we established comparison tests to simultaneously monitor the same slope with a perpendicular camera, of which the sightline was vertical to the laser planes, and a slanted camera with the sightline bias to the laser faces. The real volume of the observed slope, V_p (cm³), is calculated as

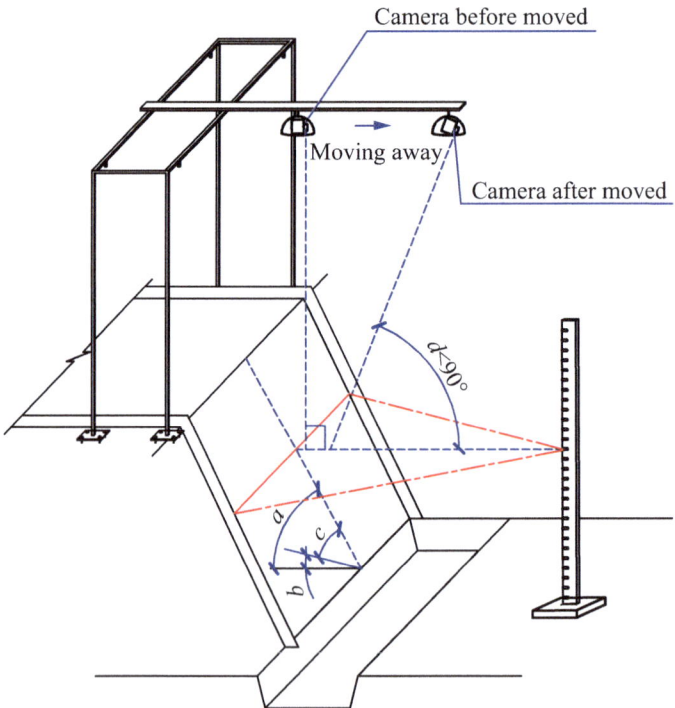

Fig. 5.8 Another way to obtain a clear video for the very steep slope. The camera is moved away from the slope. Nevertheless, in the condition, the camera sightline is not perpendicular to the laser plane, and the measured volume should be corrected

Table 5.2 Comparison of the volumes of the same slope observed with the camera from the vertical direction and that from the bias direction

Rainfall event	V_p (cm³)	V_b (cm³)	$r = V_p/V_b$
1	1,953,363.86	2,230,279.30	0.88
2	1,945,538.72	2,227,801.09	0.87
3	1,943,204.53	2,224,054.85	0.87
4	*	–	–
5	1,954,963.61	2,220,987.19	0.88
6	1,953,627.73	2,217,715.87	0.88
Average	–	–	0.88

The data were obtained from the experimental group, 3-FL60-1-60d, of which the rainfall intensity was 0.8 mm/min and the duration was 60 min, and the slope height and gradient of the initial landform before the first rainfall were 1 m and 60°, respectively. The experiments were conducted in August, 2014. Six runs of rainfalls, each at an amount of 48 mm, were applied in turn
Notes V_p is the volume measured with the camera of which sightline is perpendicular to the laser faces, and V_b is the volume measured with another camera of which sightline is bias to the laser faces
*The data was missed due to mis-operation

follows: $V_p = V_b \times r$, where V_b is the volume observed with the slanted camera (cm³) and r is the ratio of the slope volume obtained from the perpendicular camera to that from the slant one. Figure 5.8 shows the topography images that were captured from the camera with the sightline perpendicular to the laser faces and another camera with the sightline bias to the laser faces. These images were all obtained from the second rainfall of the Experiment 3-FL60-1-60d. In the experiment, the rainfall intensity was 0.8 mm/min and the rainfall duration was 60 min; the slope height and the slope gradient of the initial landform before the first rainfall were 1 m and 60°, respectively. The white boundaries in the two figures were drawn according to the same control points on the flank wall. In the second and third rows of Table 5.2, the volumes of the slope surrounded by a white frame (i.e., V_p and V_b) were observed from the camera of which sightline was perpendicular to the laser faces and from another camera with sightline bias to the laser faces with an angle of 80°. As a result, the ratio, r, of V_p to V_b is close to a constant, 0.88. The projected area of the slope enclosed by the white frame in Fig. 5.9b, which was observed with the slanted camera, also seems obviously larger than that in Fig. 5.9a, which was observed with the perpendicular camera.

The first method shown in Fig. 5.7 is preferred to be adopted, for the image captured with the camera is not distorted, and the volume calculated according to this method is the real one. In contrast, if we use the second method shown in Fig. 5.8, the slope volume must be corrected because of the image distortion. However, the method could effectively enhance the definition of the figure captured with the slanted camera. In practice, a combination of the two methods is frequently used.

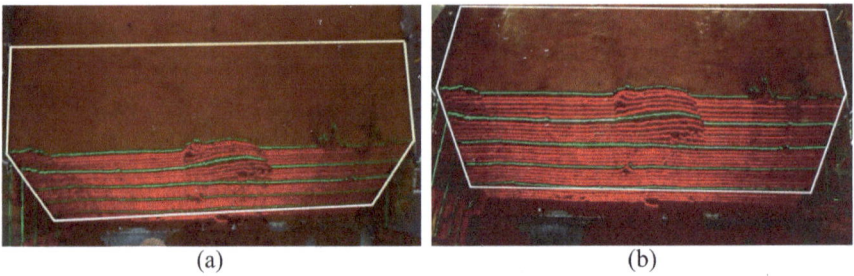

(a)   (b)

Fig. 5.9 A slope observed in the cameras from different directions. **a**, **b** show the images captured from the camera with sight line perpendicular to the laser face and another camera with sight line bias to the laser face, respectively. Volumes of the slopes surrounded by a white frame in **a**, **b** are the same

5.3.3 Assessment of the MX-2010-G Topography Meter

All results seem rational and reliable. The MX-2010-G topography meter could quantitatively interpret the process of mass movement on the steep loess slope under rainfall simulation. The ability to capture cubic centimeter accuracy data at high resolution and the synchronous capability afforded by the topography meter, make the topography meter the ideal tool for quantitative mapping of soil erosion. During the period of the experiment, the movable laboratory had stood the test of severe weather, of which rainfall in 24 h was more than 100 mm and wind force was over 10 grades. Compared with the indoor laboratory, a similar observation environment has been created in our mobile laboratory on the loess slope. In other words, in our mobile laboratory, measurement accuracies of artificial simulation rainfall, gravity erosion, runoff, and sediment yield have reached the level of those in laboratory, so that the obtained results can be compared with earlier published gravity erosion results of the indoor laboratory. No analogous approach has been found in the existing literature to date. The measurement system was used to complete the survey for over 130 rainfall simulation events; it confirmed the feasibility and reliability of this technique.

The video image stored in the memory of the computer is an excellent reference to deeply analyze the slope behavior and interactivity between images and measurement results—this can be done only by "clicking" onto the image of the object shown on the screen of the notebook. Alternatively, because the slope behavior had been recorded in the video image, data could be stored in computer memory, and landform could be monitored in either a static mode or a kinematic mode. Not only the positions were computed, but also the image of the ground was generated. Moreover, the recorded image was especially pivotal to the gravity erosion. The proposed topography meter had the advantage that the map could be checked in progress for omissions and reliability.

The topography meter was a rugged instrument, comparatively easy to operate, and could be conveniently adjusted in the field. In the fieldwork, an electricity generator was used to provide a constant power supply to the equipment.

However, it is worth noting that a more intelligently steered instrument is desired. The procedures for calculating the volume of soil loss involve many separate pieces of software and subjective decisions to distinguish whether the soil loss is caused by hydraulic erosion or gravity erosion. Hence, a surveyor's skills are still essential for precise measuring for most surveying tasks. One of the directions for future research and development of the topography meter should focus on achieving capable systems for even an unskilled surveyor to obtain the required accuracy level as well as proper/efficient data collection or setup, including post-processing for any type of measurement, at the same speed as the work done by a skilled surveyor.

In this study, when the topography meter was carried to the field to conduct the site survey, the system should be adjusted to make the lasers parallel and equidistant before the commencement of the slope behavior survey. However, the work is a bit laborious, and it also reflects a shortcoming of the topography meter, because adjusting the internal structure of an instrument should be undertaken by the manufacturer rather than the user. Presently we are developing a new type of topography meter, of which the module positions are immobilized. The laser source will consist of several units. Every unit is a box filled with a kind of rapid hardening resin and screwed on the mainboard. The laser module is put into the box and surrounded with the resin. Before the resin hardens, the position of the module could be modified. Nevertheless, when modules in the boxes were adjusted to be parallel and equidistant, we could make the resin concretionary. If a module malfunctions, the module should be replaced.

5.4 Conclusions

This chapter presents a novel topography meter based on a structural laser together with a movable tent to observe the random gravity erosions in the field. The performance of the proposed measurement system has been demonstrated in the field landslide experiments. The experimental results indicate that the apparatus was at a sufficiently high accuracy, enough for the mechanism exploration of slope erosion. The real-time digital videos of the measured surroundings recorded in the topography meter are also of inherently high value as a proof for further research. As a result, the proposed measurement system provides an ideal solution for a wide range of landform inspection and soil loss measurement, especially for monitoring gravity erosion on steep slopes. In the near future, a more intelligently steered instrument is desired that allows an unskilled operator to obtain efficient and accurate results at the same level a s hose obtained by a skilled operator.

References

Casula G, Mora P, Bianchi M G. 2010. Detection of terrain morphologic features using GPS, TLS, and land surveys: "Tana della Volpe" blind valley case study. Journal of Surveying Engineering, 136(3): 132–138.

Gokceoglu C, Sezer E. 2009. A statistical assessment on international landslide literature (1945–2008). Landslides, 6(4): 345–351.

Hergarten S. 2012. Topography-based modeling of large rockfalls and application to hazard assessment. Geophysical Research Letters, 39(13): L13402.

MBMRYR (Management Board for the Middle Reaches of the Yellow River of the Yellow River Conservancy Commission). 1993. The 8th volume of the Yellow River Annals: Recapitulation of soil and water conservation in the Yellow River watershed. Zhengzhou: Henan People's Publishing House (in Chinese).

Xu C, Xu X W, Shyu J B H, et al. 2015. Landslides triggered by the 20 April 2013 Lushan, China, Mw 6.6 earthquake from field investigations and preliminary analyses. Landslides, 12(2): 365–385.

Zhang C, Wang D G, Wang G L, et al. 2014. Regional differences in hydrological response to canopy interception schemes in a land surface model. Hydrological Processes, 28(4): 2499–2508.

Zhang A J, Zhang C, Chu J G, et al. 2015. Human-induced runoff change in Northeast China. Journal of Hydrologic Engineering, 20(5): 04014069.

Part II
Soil Conservation Experiments: Case Studies on the Loess Plateau, China

The work develops new data sets to simulate the geomorphic process of slope land, especially on the steep slopes, and quantify various forms of soil loss, including hydraulic erosion and gravity erosion.

Chapter 6
A Close Look of the Gravity Erosion on the Loess Plateau of China

Abstracts Landslide plays an important role in landscape evolution, delivers huge amounts of sediment to rivers and seriously affects the structure and function of ecosystems and society. Here, a statistical analysis together with a field investigation was carried out on the Loess Plateau of China. Since the 1980s, 53 fatal landslides have occurred, causing 717 deaths. As the most important trigger, rainfall induced 40% of the catastrophic landslides, while other factors, i.e., human activities, freeze-thaw and earthquake, accounted for 36, 23 and 1%, respectively. Furthermore, the landslide frequency and death toll related to human activities were increasing as time went on. Landslide also plays an important role in sediment delivery, especially in the areas with steep terrain. In some catchments of the Loess Plateau, landslides contributed over 50% of the total sediment discharge. The result shows that landslide is a widespread geologic hazard in the rural area of the Loess Plateau, China.

Keywords Geological hazards · Landslides · Soil yield · Triggering mechanism

6.1 Gravity Erosion: Natural Hazard and Soil Erosion

A landslide is the downslope movement of soil, rock and organic materials under the effects of gravity and also the landform that results from such movement (Highland and Bobrowsky 2008). Serious landslides occur in nearly all areas of the world, and they could cause serious natural hazards and major ecological and environmental issues such as soil erosion on steep slopes, billions of dollars of damage and many casualties (Zhang et al. 2014; Keefer and Larsen 2007). In the period from 2004 to 2010, there were 2620 fatal landslides worldwide, resulting in 32,322 deaths (Qiu 2014). For instance, a landslide in the southern Philippines led to 122 deaths and 1328 missing (Stone 2006). Landslides are also especially prevalent in the rural area of the Loess Plateau of China. According to investigations and statistics, landslides have resulted in approximately 1000 fatalities per year over the past 20 years in China (Huang 2009), and one-third of them occurred on the Loess Plateau (Zhuang and Peng 2014). For example, catastrophic landslides on the Loess Plateau occurred on March 7, 1983 in Dongxiang County, September 17, 2011 in Xi'an City and August

© Science Press and Springer Nature Singapore Pte Ltd. 2020
X. Xu et al., *Experimental Erosion*,
https://doi.org/10.1007/978-981-15-3801-8_6

12, 2015 in Shanyang County, resulting in 237, 32 and 65 casualties, respectively (CGEIS 2016; Huang 2009). In addition to the loss of human life and property directly due to landslides, the corresponding sedimentation is another significant issue that deserves attention (Tsai et al. 2012). For example, in the headwaters of the 2150 km^2 Waipaoa River catchment, an area well known for its severe erosion in New Zealand, gullies contributed 59% of the total sediment load. The amount of slumps, shallow riparian landslides and earthflows accounted for 34, 6 and 1% of the total sediment load, respectively (Marden et al. 2014). On the Loess Plateau of China, landslide plays a more important role in yielding sediment in the small watershed. Hence, how to prevent or mitigate disasters triggered by landslides is an urgent problem.

The Occurrence of landslide is controlled both by a series of internal factors, e.g., geological, geomorphological, soil property and hydro-geological and external factors, including seismicity, heavy rainfall, human activities and freeze-thaw (Conforti et al. 2015; Zhuang and Peng 2014). Most landslides occur during or just after storms. In China, 50% of large-scale landslides with volume greater than two million m^3 that occurred in the past three decades were induced directly by rainfalls (Huang 2009). In addition, adverse destabilization from human activities has been another major contributory factor in triggering landslides since the 1980s in China. With the development of the Chinese economy, residential areas on the Loess Plateau being rapidly expanding along steep slope terraces and adverse destabilization from human activities; i.e., no protection practice after slope cutting, is the major cause of landslide incidents (Xu et al. 2015a). The steep-cut slopes encourage the concentration of shear stress at the foot of the slope and tension stress at the top, leading to the formation of cracks in the inner slope (Zhuang and Peng 2014), which results in the instability of the side slopes. Moreover, seismic activities make the slopes unstable and more vulnerable to failures. In 2008, the Wenchuan earthquake in China triggered a vast number of landslides (Parker et al. 2011). Finally, freeze-thawing is an influential factor in triggering landslides. In Northwest China, thawing of the frozen layer in spring is responsible for triggering some landslides in the loess region. For instance, a landslide in Dongxiang County killed 237 people (Huang 2009). However, the different combinations of these geo-environmental factors lead to differences in mechanisms of landslide destruction. Evaluating the behaviors and triggers of landslides has significant implications for understanding their kinematic mechanisms and evolution.

Field investigation and statistical analysis are among the most important tools for recognizing and evaluating the landslides. Field investigations can provide a basis for better understanding and illustrating the spatial distributions and hazards related to landslides (Xu et al. 2015b). In recent years, field investigations were always immediately performed after major landslides in China, such as the landslides in Wenchuan County triggered by the earthquake in 2008 (Yang et al. 2015) and landslides in Guanling County triggered by rainfall in 2010 (Xing et al. 2014). The results of these investigations provided useful information for landslide mitigation and detailed landslide inventory compilation. A field observation in the Guanzhong area of the Loess Plateau indicated that the main type of landslide was deep-seated rotational landslides with sliding masses composed of two different packages of strata (Li et al. 2015). Another field investigation of 13 loess landslides in Yan'an area of the Loess Plateau showed that the infiltration depth of rainfall in integrated loess mass is generally limited to 3.0 m underground, and rainfall infiltration inducing loess landslide can be divided into three modes (Tang et al. 2015). Through the field investigation and statistical analysis based on the 52 loess avalanches in the Baota district of Yan'an, Qiu et al. (2015) found power law correlations between avalanche volume and avalanche area. The other field investigations also showed although the avalanche in Tuban was small in size, the death toll and financial losses were shocking (Xu et al. 2015a). All these studies focused on location, type and modes of the landslide and disaster losses. A commonly used method to estimate the sediment yield of existing landslides is to calculate the change in surface elevation over time based on digital elevation models (Tsai et al. 2012). Nevertheless, few papers have examined the role of soil erosion caused by landslides, especially those in small size.

For years, the authors of this study have carried out field investigations and statistical analyses of landslides on the Loess Plateau of China. The objectives of this study are to analyze hazards and soil erosion caused by the loess landslides and then to determine how t o control the hazards that occur so widely in the rural areas of the Loess Plateau.

6.2 Characteristics of the Loess Plateau

The Loess Plateau is located in the upper and middle reaches of the Yellow River, covering a total area of 624,000 km^2, and over 60% of the land is subjected to serious soil and water loss. It is mostly located in arid or semiarid regions, featuring lots of dry air, little cloud and abundant sunlight. The average annual precipitation on the Loess Plateau is only 350–550 mm, decreasing gradually from the southeast to the northwest, and most of the precipitation is concentrated in the rainy season from June to September (Xu et al. 2004). Usually, a single rainstorm, most of which is in the form of short but very intense rainfalls, can account for more than 60%, even 90%, of the total precipitation in the year, and causing the greatest amount of soil loss (Zhou et al. 2000). The rain often lasts 30–120 min, which will result in a 10–40 min hyperconcentrated flow with an average concentration of 200–300 kg/m^3 and a maximum suspended concentration of 1000 kg/m^3 (Wang and Jiao 1996).

The most crucial area is the Loess Mesa Ravine Region and the Loess Hill Ravine Region, which cover altogether 30% of the total area. As one of the most severe areas of soil and water losses in the world, the average erosion rate can be as high as 5000–10,000 t/(km^2 a), and sometimes even up to 20,000–30,000 t/(km^2 a) (Meng 1996). In the area, landslides frequently occur that are induced by the rainstorm, slope excavation, freeze-thaw and earthquake, because the undulating terrain is very steep, the vegetation is so sparse, and especially because the loess is collapsible and in vertical joints.

6.3   Methods and Materials

Most of the landslide data were collected from the Bulletin of National Geological Disaster (CGEIS 2016), and others were obtained from academic monographs, journal articles and the investigation undertaken by the authors of this book. In the Bulletin of National Geological Disaster, the location, time, volume, death toll and direct economic loss of each landslide were collected in field investigations conducted by the expert group, and triggering factors were analyzed according to the results of field investigation together with the data on meteorology, precipitation, earthquake and engineering. The landslide volumes were obtained from the field measurement or the field investigation coupled with remote sensing interpretation or on-site survey combined with aerial photograph and synthetic aperture radar. In the academic monographs and journal articles, landslide data were gathered from the field investigation, exploratory boring, monitoring, and survey of the representative landslides and interpretation of remote sensing results. For instance, Tang (2014) performed field investigation in the southern area of Yulin in the northwest of China to obtain the loess landslide data about the location, time, volume and triggering factors, and so did Song et al. (1994). Wang (2004) and Liu (1998) analyzed triggering factors and creep behavior via field investigation and deformation monitoring, while Guo et al. (2014) obtained the amount of landslide through field measurement coupled with remote sensing interpretation. Furthermore, a geophysical survey was chosen to estimate the landslide volume (Huang 2009), whereas a comprehensive method including ground survey, measurement of typical landslides and remote sensing interpretation was utilized to acquire landslide data about geometrical characteristics (Chen et al. 2013).

To get a close look at the landslides on the Loess Plateau, the authors trudged along the Liudaogou Catchment in Shenmu County of Shaanxi Province from July to August, 2014. After that, the authors also trekked into the rural area of the Lvliang City, Shanxi Province in May, 2015 to investigate local disasters, including landslides in Tuban Town, Houwa Village, Shang'an Village and Wangjiagou Catchment, as well as some small-scale mass failures in other areas. All these collapse sites were measured and photographed. In addition, the authors interviewed local villagers and the researchers in the Water Conservancy Bureau of the local government. The distances between the control points were measured with a long steel tape and a laser range finder. Subsequently, pictures of the landslides containing the control points were put into AutoCAD to portray the margin of the landslides, and then the collapse area within the margin could be calculated in the software. The volumes of undisturbed failure blocks could be directly measured through the laser range finder and the tape. However, for the landslide sites that had been cleared of residual materials, the volumes of the failure blocks could not be obtained by direct calculation. Fortunately, from previous experiences reported in the literature, the empirical relationship linking landslide area A and volume V could be used to calculate the volumes of the failure blocks (Chen et al. 2015; Guzzetti et al. 2008). Here, the volume-area scaling relation is used to estimate individual landslide volumes: $V = 6.4473 \times A^{1.1422}$, where V is the landslide volume (m^3) and A is the landslide area (m^2) (Qiu et al. 2015). This relation, of which the correlation coefficient was 0.91 and the significance level was less than 0.05, was developed from a catalogue of 52 landslides surveyed in Yan'an area, Shaanxi Province. Both the Lvliang area investigated in this study and the Yan'an area are located in the Loess Hill Ravine Region, and the average thickness value of the individual landslides in Lvliang section is close to that in Yan'an. Hence, the volume-area scaling relation can be employed to calculate the individual landslide volumes of disturbed failure blocks. The sediment yield is known as a net soil loss, which is calculated by subtracting in-field deposition from gross soil loss (Tsai et al. 2012).

According to the failure volumes, death tolls and economic losses of the landslides, the Ministry of Land and Resources of the People's Republic of China (2004) divided the landslide disasters into four grades: small-scale loss event, moderate loss event, severe catastrophe and devastating catastrophe. The landslides analyzed here were the same as or more serious than a moderate event with a collapsed volume larger than 10,000 m^3 and/or death toll less than 1. In this study, the effects of rainfall, human activities, freeze-thaw and earthquake on the occurrence of the landslide are calculated by the equal-weight method when one landslide event was preceded by two or more kinds of triggers. For instance, if a landslide was triggered only by rainfall, the weighting coefficient of the factors is 1; if a landslide was induced by rainfall events and human activities together, both weighting coefficients of the two factors are assumed to be 0.5.

6.4 Results and Discussion

6.4.1 Analysis of the Catastrophic Loess Landslides

The catastrophic landslides that happened on the Loess Plateau from 1980 to 2015 were chosen for a detailed case study (Table 6.1). All the accidents were beyond moderate events and had definite records of date and location in official documents. In total, there were 53 fatal landslides, causing 717 casualties. The proportion of landslide induced by each factor could be found via a statistical analysis of the data. It turned out that rainfall, inducing 40% of the catastrophic events, was the most important trigger factor for initiating catastrophic landslides. The result was particularly linked to the intense and concentrated summer rainfalls on the Loess Plateau. For example, the landslides in the Baiyuli Village and the Wangjiagou Village of Shanxi Province and the debris flow in the Min County of Gansu Province were all directly triggered by storms. On the other hand, landslides induced by the earthquake and freeze-thaw process accounted for 1 and 23% of the total, respectively. Hence, the process of freeze-thaw has a greater impact on landslides than earthquakes.

Table 6.1 Catastrophic landslides on the Loess Plateau since 1980s

Location	Reference	Occurrence date	Triggering factor	Volume (10^3 m^3)	Death toll	Direct economic loss (10^6 Yuan RMB)
Zhongcun Town, Shanyang County, Shaanxi Province	CGEIS (2016)	August 12, 2015	Mining and rainfall	1680.0	65	–
Tuban Town, Lvliang City, Shanxi Province	*	May 19, 2015	Excavation	5.0	7	–
Houwa, Lvliang City, Shanxi Province	*	July 15, 2013	Rainfall and excavation	5.6	4	–
Zhangjiazui Village, Zhongyang County, Shanxi Province	CGEIS (2016)	November 16, 2009	Rainfall and snow	25.0	23	0.2
Eryang Gully, Min County, Gansu Province	Guo et al. (2014) and CGEIS (2016)	May 10, 2012	Rainfall	2100.0	9	130
Shijiadao Village, Xi'an City, Shaanxi Province	CGEIS (2016)	September 17, 2011	Slope cutting and rainfall	240.0	32	52
Jialing River, Nveyang County, Shaanxi Province	CGEIS (2016)	July 5, 2011	Earthquake and rainfall	5.0	18	10
Baiyuli Village, Dai County, Shanxi Province	CGEIS (2016)	June 26, 2011	Rainfall and mining	52.2	9	–
Saerta Square, Dong County, Gansu Province	CGEIS (2016)	March 2, 2011	Freeze-thaw	180.0	0	443

(continued)

Table 6.1 (continued)

Location	Reference	Occurrence date	Triggering factor	Volume (10^3 m^3)	Death toll	Direct economic loss (10^6 Yuan RMB)
Donghua Town, Huating County, Gansu Province	CGEIS (2016)	July 24, 2010	Rainfall	0.8	13	0.16
Gaoba Town, Shanyang County, Shaanxi Province	CGEIS (2016)	July 24, 2010	Rainfall	250.0	24	1.5
Siji Town, Ankang City, Shaanxi Province	CGEIS (2016)	July 18, 2010	Rainfall	20.0	20	–
Dazhuyuan Town, Ankang City, Shaanxi Province	CGEIS (2016)	July 18, 2010	Rainfall	400.0	29	–
Shuanghuyu Town, Zizhou County, Shaanxi Province	Tang (2014)	March 10, 2010	Freeze-thaw	89.0	27	–
Shixiakou District, Lanzhou City, Gansu Province	CGEIS (2016)	May 16, 2009	Slope cutting and rainfall	30.0	7	20.6
Yanjiahe Village, Zizhou County, Shaanxi Province	Tang (2014)	November 27, 2008	Cave dwelling	1.5	2	–
Xuecha Town, Wuqi County, Shaanxi Province	Chen et al. (2013)	August 28, 2008	Slope cutting	66.0	0	–
Shang'an Village, Lvliang City, Shanxi Province	*And CGEIS (2016)	June 13, 2008	Slope cutting and irrigating	10.0	19	3

(continued)

Table 6.1 (continued)

Location	Reference	Occurrence date	Triggering factor	Volume (10^3 m^3)	Death toll	Direct economic loss (10^6 Yuan RMB)
Jiuheping Village, Zizhou County, Shaanxi Province	Tang (2014)	February 27, 2008	Freeze-thaw	4.8	5	–
Kuantan Village, Suide County, Shaanxi Province	Tang (2014)	October 12, 2007	Rainfall	0.4	4	–
Haojiapan Village, Suide County, Shaanxi Province	Tang (2014)	July 13, 2007	Rainfall	3.0	1	–
Xiahua Town, Hejin City, Shanxi Province	CGEIS (2016)	April 22, 2007	Rainfall excavation	5.0	12	5
Daming Town, Hua County, Shaanxi Province	CGEIS (2016)	October 6, 2006	Canal leakage and rainfall	55.0	12	1
Jichang Town, Ji County, Shanxi Province	CGEIS (2016)	May 9, 2005	Cave dwelling and rainfall	600.0	24	20
Shizhaihe Village, Zizhou County, Shaanxi Province	Tang (2014)	December 13, 2004	Excavation	1.2	4	–
Xujiagou Village, Zizhou County, Shaanxi Province	Tang (2014)	October 1, 2004	Rainfall	14.0	1	–

(continued)

Table 6.1 (continued)

Location	Reference	Occurrence date	Triggering factor	Volume (10^3 m^3)	Death toll	Direct economic loss (10^6 Yuan RMB)
Dangjiagou Village, Suide County, Shaanxi Province	Tang (2014)	May 27, 2004	Excavation	1.5	2	–
Wujiagou, Zizhou County, Shaanxi Province	Tang (2014)	March 13, 2004	Freeze-thaw	0.06	1	–
Liangjiagou Village, Zizhou County, Shaanxi Province	Tang (2014)	September 23, 2003	Rainfall	108.0	12	–
Xindian Village, Suide County, Shaanxi Province	Tang (2014)	July 21, 2003	Excavation, rainfall	15.0	1	–
Shipanhe Village, Zizhou County, Shaanxi Province	Tang (2014)	January 21, 2003	Snow and frost	36.0	4	–
Luoyuan Subdistrict, Wuqi County, Shaanxi Province	Chen et al. (2013)	July 25, 2002	Slope cutting	0.6	17	–
Laojundian Town, Zizhou County, Shaanxi Province	Tang (2014)	November 8, 2001	Rainfall	14.0	0	–
Gaojiaping Town, Zizhou County, Shaanxi Province	Tang (2014)	January 11, 2000	Reservoir leakage	48.0	2	–

(continued)

Table 6.1 (continued)

Location	Reference	Occurrence date	Triggering factor	Volume (10^3 m^3)	Death toll	Direct economic loss (10^6 Yuan RMB)
Changqingqiao Town, Ning County, Gansu Province	Wang (2004)	April 29, 1998	Rainfall	1.0	8	–
Huangjiagou Village, Suide County, Shaanxi Province	Tang (2014)	August 4, 1997	Rainfall	120.0	5	–
312 State Road, Huining County, Gansu Province	Wang (2004)	July 15, 1996	Rainfall	200.0	4	–
Baizigou Coal Mine, Bin County, Shaanxi Province	Wang (2004) and Liu (1998)	July 6, 1995	Mining	1800.0	0	–
Yanguoxia Town, Yongjing County, Gansu Province	Wang (2004)	January 31, 1995	Irrigation	6000.0	0	–
Rangou Village, Wubao County, Shaanxi Province	Tang (2014)	November 20, 1993	Freeze-thaw	6.0	6	–
Xindian Town, Suide County, Shaanxi Province	Tang (2014)	February 4, 1990	Snow and cave dwelling	100.0	6	–
Baita Mountain, Lanzhou City, Gansu Province	Chen et al. (2013)	November 9, 1986	Irrigation	2.5	7	–
Xujiawa, Lanzhou City, Gansu Province	Chang (2013)	June 28, 1986	Rainfall	60.0	0	–

(continued)

Table 6.1 (continued)

Location	Reference	Occurrence date	Triggering factor	Volume (10^3 m^3)	Death toll	Direct economic loss (10^6 Yuan RMB)
Hejiagou Village, Qingjian County, Shaanxi Province	Tang (2014)	May 26, 1985	Rainfall	0.8	9	–
Liujiagou Village, Jia County, Shaanxi Province	Tang (2014)	January 16, 1985	Freeze-thaw	70.0	5	–
Penzegou Village, Qingjian County, Shaanxi Province	Tang (2014)	January 8, 1985	Freeze-thaw	1.0	2	–
Guliu Village, Xi'an City, Shaanxi Province	Wang (2004) and Song et al. (1994)	December 15, 1984	Rainfall and reservoir leakage	3100.0	0	–
Sale Mountain, Dongxiang County, Gansu Province	Huang (2009)	March 7, 1983	Freeze-thaw	31,000.0	237	–
Wujiagou Village, Zizhou County, Shaanxi Province	Tang (2014)	April 30, 1982	Freeze-thaw	720.0	7	–

The samples were selected from those that occurred on the Loess Plateau during 1980–2015, which had definite records of date and location. Moreover, the landslides listed here are the same as or more serious than those belonging to moderate events. That is to say, the volume of an individual landslide was equal to or larger than 10^4 m^3 and/or the death toll was not less than 1

*Investigated by the authors of this study on May, 2015

   In addition to the rainfall factor, adverse destabilization from human activities has become another major contributory factor to landslides since the reform and opening up of China. Inferred from statistical data mentioned above, the number of direct and indirect human-induced landslides accounted for 36% of the total. As shown in Fig. 6.1, the landslide frequency and death toll related to human activities were also increasing over time. The quantities of catastrophic landslides and deaths caused by human activities occurring in five years from 2010 to 2015 were 4 and 28 times than those during the period 1980–1990, respectively. Investigations by the first author of this study show that the avalanche in Tuban on May 19, 2015, as shown in Fig. 6.2a, was mainly caused by human excavation and might have been directly triggered by unrecorded slight geological activity (Xu et al. 2015a). Similarly, excavation at the slope toe was the major reason to cause the failure in Houwa (Fig. 6.2b), while slope cutting to get soil was a big contribution to the collapse in Shang'an (Fig. 6.2c). In the Houwa Village, as the residents cut a terrace to build a house near the loess slope, the toe of slope was damaged, the lateral pressure on the slope was lost, and then the slope ultimately became unstable. Moreover, a continuous heavy rainfall for eight days occurred before failure (Weather Network 2013), increasing the sliding force and decreasing the shear strength of loess. Finally, the landslide happened through the sliding zone in which the shear strength of the loess was insufficient to resist the shear stresses. Continuous excavation led to the collapse in Jiuxing Brickyard, Shang'an Village. The brickyard had cut slope to obtain soil for ten years. As a result, a high and steep slope with a gradient of 70°–90° and a large area of the free surface was formed, and stability of the side slope was seriously weakened. On the other hand, before the landslide, planting and watering near the slope further contributed to more instability of the slope (Xinhuanet 2008). The occurrences of mass failures in Tuban, Houwa and Shang'an were closely related to human activities.

Fig. 6.1 Frequency and death toll of landslides caused by human activity from 1980 to 2015

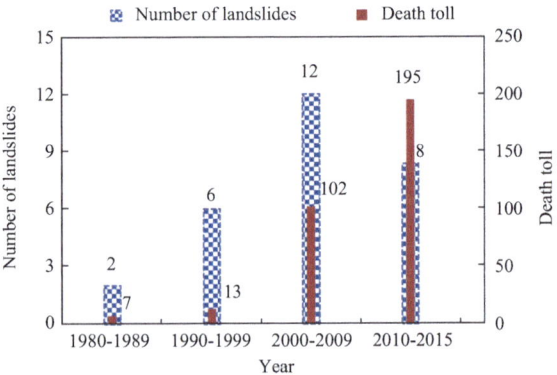

Fig. 6.2 Onsite views of representative landslides on the Loess Plateau. **a** An avalanche in Tuban Town; **b** A landslide in Houwa Village; **c** An avalanche in Shang'an Village

6.4.2 Soil Erosion Caused by Loess Landslides

Landslide debris in mountainous areas is the major source of sediment delivered to downstream areas and dominates the mountain erosion rates (Lin and Chen 2012; Chuang et al. 2009; Dadson et al. 2004). Landslide is widespread in the study area and plays an important role in the geomorphic evolution followed by vegetation deterioration and ecological damages. Field observations attribute the growth of headwaters and the widening of many gully systems to landslide erosion (Fig. 6.3). On the relatively steep slope, landslide in association with gullies formed gully-landslide complexes, which represented the main kind of geo-environmental degradation in the area. Landslides gave rise to the uncovered soil surface where vegetation recovered slowly. Figure 6.3b shows a failure scar in the Wangjiagou Catchment caused by a landslide in July, 2013. Plants on the slump surface were still very sparse.

Site inspection indicated that lots of small landslides happened in rural areas of the Loess Plateau. Far from the residential area, landslides did not cause casualties

(a)	(b)

Fig. 6.3 Plant deterioration and soil loss caused by landslide erosion on the steep slope. **a** An active slope shaped by landslide erosion in the Liudaogou Catchment, Shaanxi Province; **b** A failure scar in the Wangjiagou Catchment, Shanxi Province

or economical loss, but they always led to serious soil loss. Although these failures were typically only a few meters deep, they covered extensive slope areas and produced great volumes of colluvium that filled valley bottoms and locally blocked floodplains, which are always the main source of hyper-concentration flow when they experience subsequent rainstorms and floods. It was indicated that the landslide erosion made sediment concentration increase up to 1000 kg/m^3, leading to hyper-concentrated flow (Li et al. 2009). Moreover, the landslide erosion was so serious that its contribution to total erosion was over 50% in some catchments on the Loess Plateau (Table 6.2).

Monitoring the amount of soil loss caused by small landslides is very difficult because of the randomness and suddenness of such events. Some references reported the role of erosion caused by mass failure. As shown in Table 6.2, R is expressed as the ratio in percentage between the sediment discharge from landslides and the total soil loss of a catchment. Jiang et al. (1966) estimated that the ratio R was up to 72% in the Nanxiaohegou Catchment of which area was 31 km^2 on the Loess Mesa Ravine Region, while R was only 20% in the Jiuyuangou Catchment of which area was 70 km^2 on the Loess Hill Ravine Region. However, the sediment discharge from landslides was obtained only through integration of the historical data from runoff plots and hydrologic station in the catchment, and no quantitative observation method or theoretical calculation method was found in this paper. In recent years, some theoretical calculation methods have been utilized to determine the ratio R by using the observed hydrologic data and sediment data from the tributaries of the Yellow River. These methods include the dynamic model of river sediment transport (Wang and Li 2009; Li et al. 2009) and the sediment transport capacity of overland flow (Yang et al. 2014). Wang and Li (2009) and Li et al. (2009) reported that in the Chabagou Watershed with an area of 205 km^2 and the Wudinghe Watershed with an area of 29,600 km^2, the ratios were 21 and 12%, respectively. Yang et al. (2014) found that R was up to 42% in the Qiaogou Catchment with an area of 0.45 km^2. Nevertheless, such estimates are often based on theoretical calculations, while the lack of additional observed data places constraints on the ability to verify accuracy. Researchers also investigated the location, time, scale and distribution of landslides

Table 6.2 The role of landslide erosion in the upper and middle reaches of the Yellow River

Catchment/watershed	Reference	Statistical time	Area (km^2)	Landslide volume (10^3 t)	*R (%)	Averaged R (%)
Qiaogou Catchment	Yang et al. (2014)	2009–2011	0.45	0.65	42	39
Luoyugou Watershed	TSSWS (2004)	1987	73	446	12	
Lv'er Catchment	TSSWS (2004)	1961	12	100	50	
Nanxiaohegou Catchment	Jiang et al. (1966)	1954–1960	31	86	72	
Jiuyuangou Catchment	Jiang et al. (1966)	1954–1960	70	241	20	
Chabagou Watershed	Li et al. (2009), Wang and Li (2009)	1967	205	467	21	17
Wudinghe Watershed	Wang and Li (2009)	1977	29,600	5230	12	

*R is expressed as the ratio in percentage between the sediment discharge from landslides and the total soil loss of a catchment

in the small watershed and then estimated the amount of the landslide erosion by measuring the volume of soil loss from the failure surface and the landslide deposit; for example, TSSWS (2004) reported the contribution of landslide erosion to the total sediment discharge with this method in the Lv'ergou Catchment and the Luoyugou Catchment. In fact, the upstream sediment-laden runoff might take away part of the landslide deposit or result in a large quantity of sediment deposition. Hence, errors existed as the landslide volume was calculated in such a way according to the state of failure scar after rainfall.

All in all, landslide erosion accounts for a considerable proportion of soil erosion in the middle reaches of the Yellow River, and it is also an important contributor to sediment input to the Yellow River. Table 6.2 shows that the proportion of land-slide erosion to total erosion generally decreased, while the area of the watershed increased. In fact, the detached soil from gully banks could enter the gully runoff directly, because the landslide erosion was mainly distributed in the gully-head and branch gully. However, for large-scale channels, the ratio of width to depth was relatively large, so that detached soil could not directly get into runoff even if landslide erosion occurred. On the other hand, as runoff grew with increased spatial scale, the detached soil in a single landslide event could not significantly affect the total sediment concentration (Li et al. 2009). In Table 6.2, the landslide erosion accounted for 39% of the total erosion at the catchment scale with an area less than 100 km^2. However, the average ratio of landslide erosion to the total erosion was only 17%

in the Chabagou Watershed with an area of 205 km^2 and the Wudinghe Watershed, which has an area of 29,600 km^2.

6.4.3   *Urgent Desires and Effective Measures to Prevent Landslides*

Due to the frequent occurrence of landslide disasters, effective measures advocated by the local government are urgently desired to control landslide hazards. As the authors were investigating the event in the Houwa Village, a woman villager burst into a storm of abuse and tried to prevent us from the accident sites for she regarded us as the officials from the government (Fig. 6.4). Obviously, high incidences of the landslides have made great difficulties for the residents in the area. From a technology perspective, controlling landslide erosion on loess slopes is not difficult. In fact, the protection practices that are commonly used on the slope of the expressway and other public buildings in the area are safe and reliable. However, most residents in rural areas of the Loess Plateau usually build settlements near steep cliffs with no revetment (Fig. 6.5). The phenomena result in numerous landslides.

For a structurally controlled landslide to develop, comprehensive control is advocated that combines structural, vegetative and managerial measures, which ensures human safety with less investment; moreover, as part of the capacity-building efforts in landslide control, an exclusion zone is also needed at the brink of scarps, which prevents any human activity from aggravating the geologic hazard (Xu et al. 2015a). For example, while retaining walls are constructed along the side banks in descending steps, drainage ditches on the top of the slope, vegetation plantation on the slope and reinforcement with grouted rocks at the slope toe, to some degree and by necessity, are also proposed. Popularizing these measures will have a great practical impact on disaster prevention and reduction on the Loess Plateau.

Fig. 6.4 Landslide victims in the Houwa Village. The female victim in the lower section of the photo was bursting into a storm of abuse

(a)   (b)

Fig. 6.5 Rural houses with no revetment. In most areas of the Loess Plateau, especially rural regions, people usually build settlements near unprotected cliffs. **a** Houses in the suburb of Lvliang City, Shanxi Province; **b** Settlements in the Liudaogou Catchment, Shenmu County, Shaanxi Province

6.5 Conclusions

Landslide is a kind of widespread geologic hazard in the rural area of the Loess Plateau, China. Since the 1980s, 53 fatal landslides have occurred, causing 717 deaths. Landslide also accounts for a considerable proportion of soil erosion in the upper and middle reaches of the Yellow River. In some catchments of the Loess Plateau, the contribution of landslide erosion to the total erosion was up to 50%.

Among the fatal landslides on the Loess Plateau since the 1980s, about 40% were triggered by rainfall. Landslides also result from several other factors such as human activities, freeze-thaw and earthquake, which accounted for 36, 23 and 1%, respectively. Compared with the earthquake on the Loess Plateau, human activities and the freeze-thaw factor were found to be more significant in triggering a landslide. Moreover, human activities had an increasing trend in the impact of the landslide. Due to the frequent occurrence of landslide disasters, effective measures advocated by the local government are urgently desired to control landslide hazards in the area.

References

CGEIS (China Geological Environment Information Site). 2016. Bulletin of National Geological Disaster 2005: 2–5, 2006: 2–4, 2007: 3–5, 2008: 1–5, 2009: 3–16, 2010: 2–7, 2011: 2–10, 2012: 1–12, 2015: 7–8, a). (2016-02-06) [2018-12-01]. http://www.cigem.gov.cn/auto/db/explorer. html?db=1006&type=1&fd=16&fv=49&uni=0&md=15&pd=210&mdd=11&pdd=5&msd= 11&psd=5&start=0&count=20 (in Chinese).

Chang T H, Lin C L, Huang F K. 2013. Slope cracks in the safety assessment of the rainfall period. Applied Mechanics and Materials, 405–408(2): 2364–2369.

Chen J X, Wang Y Z, Song F, et al. 2013. Disaster characteristics and prevention countermeasures for the loess landslide. Beijing: Metallurgical Industry Press (in Chinese).

Chen Y C, Chang K T, Lee H Y, et al. 2015. Average landslide erosion rate at the watershed scale in southern Taiwan estimated from magnitude and frequency of rainfall. Geomorphology, 228: 756–764.

Chuang S C, Chen H, Lin G W, et al. 2009. Increase in basin sediment yield from landslides in storms following major seismic disturbance. Engineering Geology, 103(1–2): 59–65.

Conforti M, Pascale S, Sdao F. 2015. Mass movements inventory map of the Rubbio stream catchment (Basilicata – South Italy). Journal of Maps, 11(3): 454–463.

Dadson S J, Hovius N, Chen H, et al. 2004. Earthquake-triggered increase in sediment delivery from an active mountain belt. Geology, 32(8): 733–736.

Guo F Y, Meng X, M Yin N W, et al. 2014. Formation mechanism and risk assessment of debris flow of "5.10" in Eryang Gully of Minxian County, Gansu Province. Journal of Lanzhou University (Natural Sciences), 50(5): 628–632 (in Chinese).

Guzzetti F, Ardizzone F, Cardinali M, et al. 2008. Distribution of landslides in the Upper Tiber River basin, central Italy. Geomorphology, 96(1–2): 105–122.

Highland L M, Bobrowsky P. 2008. The landslide handbook-A guide to understanding landslides: Reston, Virginia, USGS Science for a changing world-Circular 1325: 3–25, 129. [2018-12-01]. http://pubs.usgs.gov/circ/1325/.

Huang R Q, 2009. Some catastrophic landslides since the twentieth century in the southwest of China. Landslides, 6(1): 69–81.

Jiang D L, Zhao C X, Chen Z L. 1966. Analysis of the source of sediment in the middle reach of the Yellow River. Acta Geographica Sinica, 32(1): 20–35 (in Chinese).

Keefer D K, Larsen M C. 2007. Assessing landslide hazards. Science, 316(5828): 1136–1138.

Li T J, Wang G Q, Xue H, et al. 2009. Soil erosion and sediment transport in the gullied Loess Plateau: scale effects and their mechanisms. Science in China Series E: Technological Sciences, 52(2): 1283–1292.

Li B, Feng Z, Wang W P. 2015. Characteristics of the Sanmen Formation clays and their relationship with loess landslides in the Guanzhong area, Shaanxi, China. Arabian Journal of Geosciences, 8(10): 7831–7843.

Lin G W, Chen H. 2012. The relationship of rainfall energy with landslides and sediment delivery. Engineering Geology, 125(1): 108–118.

Liu X Z. 1998. A Prediction of Baizigou landslide in Binxian County, Shaanxi. Journal of Catastrophology, 13(1): 54–56 (in Chinese).

Marden M, Herzig A, Basher L. 2014. Erosion process contribution to sediment yield before and after the establishment of exotic forest: Waipaoa catchment, New Zealand. Geomorphology, 226: 162–174.

Meng Q M (Ed). 1996. Soil and water conservation on the Loess Plateau. Zhengzhou: Water Resource Press of Yellow River: 75 (in Chinese).

MLR (Ministry of Land and Resources, P.R.C). 2004. Geology and mineral resources industry standard of the People's Republic of China: Standard of classification for geological disaster (DZ 0238–2004). [2018-12-01]. http://wenku.baidu.com/view/190a9afdc8d376eeaeaa313a.html. (in Chinese).

Parker R N, Densmore A L, Rosser N J, et al. 2011. Mass wasting triggered by the 2008 Wenchuan earthquake is greater than orogenic growth. Nature Geoscience, 4(7): 449–452.

Qiu J. 2014. Landslide risks rise up agenda. Nature, 511(7509): 272–273.

Qiu H J, Cao M M, Wang Y L, et al. 2015. Power law correlations of geohazards in loess hilly region. Scientia Geographica Sinica, 35(1): 107–113 (in Chinese).

Song K Q, Cui Z X, Yuan J G, et al. 1994. Creep characteristics analysis and prediction of guliu slide. Chinese Jounal of Geotechnical Engineering, 16(4): 56–64 (in Chinese).

Stone R. 2006. Disaster relief. Too late, earth scans reveal the power of a killer landslide. Science, 311(5769): 1844–1845.

Tang Y M. 2014. Risk of loess landslides: assessing, monitoring and forecasting. Beijing: Science Press (in Chinese).

Tang Y M, Xue Q, Li Z G, et al. 2015. Three modes of rainfall infiltration inducing loess landslide. Natural Hazards, 79(1): 137–150.

Tsai Z X, You G J Y, Lee H Y, et al. 2012. Use of a total station to monitor post-failure sediment yields in landslide sites of the Shihmen reservoir watershed, Taiwan. Geomorphology, 139(4): 438–451.

TSSWS (Tianshui Soil and Water Conservation Science Experimental Station of the Yellow River Water Conservancy Committee). 2004. Prototype observation and law research of soil and water loss at the third deputy district in loess hilly-gully region. Zhengzhou: The Yellow River Water Conservancy Press (in Chinese).

Wang N Q. 2004. Study on the growing laws and controlling measures for loess landslide. Chengdu: Chengdu University of Technology (in Chinese).

Wang W Z, Jiao J Y. 1996. Rainfall and erosion sediment yield on the Loess Plateau and sediment transportation in the Yellow River Basin. Beijing: Science Press: 165, 221 (in Chinese).

Wang G Q, Li T J. 2009. The dynamic model of soil erosion and sediment transport in river basins. Beijing: China Water Power Press: 65 (in Chinese).

Weather Network. 2013. Historical weather data in Dalian, China, in July 2013. (2013-08-01)[2018-12-01]. http://lishi.tianqi.com/lishi1/201307.html. (in Chinese).

Xing A G, Wang G, Yin Y P, et al. 2014. Dynamic analysis and field investigation of a fluidized landslide in Guanling, Guizhou, China. Engineering Geology, 181: 1–14.

Xinhuanet. 2008. Investigation of the loess avalanche in Lishi district, Lvliang City, Shanxi Province. (2008-06-14)[2018-12-01]. http://news.xinhuanet.com/newscenter/2008-06/14/content_8369363.html. (in Chinese).

Xu X Z, Zhang H W, Zhang O Y. 2004. Development of check-dam systems in gullies on the Loess Plateau, China. Environmental Science and Policy, 7(2): 79–86.

Xu X Z, Song G D, Liu J, et al. 2015a. Avalanche in Tuban: a hazard with no defense. Natural Hazards, 79(3): 2181–2187.

Xu C, Xu X W, Shyu J B H, et al. 2015b. Landslides triggered by the 20 April 2013 Lushan, China, Mw 6.6 earthquake from field investigations and preliminary analyses. Landslides, 12(2): 365–385.

Yang J S, Zheng M G, Yao W Y, et al. 2014. Landscape Factors of Gravity Erosion in Loess Gully. Soil and Water Conservation in China, (8): 42–45,69 (in Chinese).

Yang C W, Zhang J J, Liu F C, et al. 2015. Analysis on two typical landslide hazard phenomena in the Wenchuan earthquake by field investigations and shaking table tests. International Journal of Environmental Research and Public Health, 12(8): 9181–9198.

Zhang C, Wang D G, Wang G L, et al. 2014. Regional differences in hydrological response to canopy interception schemes in a land surface model. Hydrological Processes, 28(4): 2499–2508.

Zhou P H, Zhang X D, Tang K L. 2000. Rainfall installation of simulated soil erosion Experiment hall of the state key laboratory of soil erosion and dryland farming on Loess Plateau. Bulletin of Soil and Water Conservation, 20(4): 27–30,45 (in Chinese).

Zhuang J Q, Peng J B. 2014. A coupled slope cutting—a prolonged rainfall-induced loess landslide: a 17 October 2011 case study. Bulletin of Engineering Geology and the Environment, 73(4): 997–1011.

Chapter 7
Effects of Conservation Practices on Soil, Water, and Nutrients

Abstract To comprehensively assess the merits and demerits of the conservation practices is of great importance in further supervising the conservation strategy for the Loess Plateau. This chapter calculates the impact factors of conservation practices on soil, water, and nutrients during the period 1954–2004 in the Nanxiaohegou Catchment, a representative catchment on the Loess Plateau, China. Soil erosion and nutrient loss had been greatly mitigated through various conservation practices. About half of the total transported water and 94.8% of the total transported soil and nutrients had been locally retained in the selected catchment. The retention abilities of the characteristic conservation practices were in the following order: dam farmland > terrace farmland > forest land or grassland. Hence the check dam was the most appropriate conservation practice on the Loess Plateau. The conservation practices were more powerful in retaining sediment than in reducing runoff from the Loess Plateau.

Keywords Loess Plateau · Soil management · Soil · Runoff · Nutrient

7.1 Effects of Soil Conservation Practices

Soil erosion has been identified as one of the most significant threats to land productivity and environmental quality on the Loess Plateau of China (Fu et al. 2004). Various measures, e.g., over 100,000 check dams, numerous terrace farmlands, forest lands and grasslands, have been established to reduce soil erosion since the 1950s (Xu et al. 2004). To comprehensively assess the environmental impacts of the practices is of vital importance in further supervising the conservation strategies.

Since the practical effects varied with different conservation practices in different areas, it is essential to propose suitable conservation practices with low cost-performance ratios. Vegetation was regarded as an important means of controlling soil erosion (Castillo et al. 1997; Quinton et al. 1997). Forest management had an obvious short-term (e.g., 2–5 years) impact on stream quality at local scales, while its cumulative effect was not apparent at regional scales in the mid-Atlantic Highlands (Thornton et al. 2000). On the steep slopes, the orchards worked better than the arable lands in conserving soil in the Three Gorges Reservoir Area, China

© Science Press and Springer Nature Singapore Pte Ltd. 2020
X. Xu et al., *Experimental Erosion*,
https://doi.org/10.1007/978-981-15-3801-8_7

(Meng et al. 2001), and dense plant cover, good soil structure, high soil organic material (SOM) and biological activity were assumed to be the probable causes (Maass et al. 1988). While on the steep hillside farm, the bench terrace was revealed to play a more significant role in reducing soil and water loss compared to conventional cultivation methods (Mills et al. 1992). In the Hekou-Longmen Region of the Loess Plateau, check dam was of vital importance in retaining soil in all of the conservation measures (Ran et al. 2002, 2004).

Erosion studies carried out at the catchment scale are more representative of the natural processes than those at the plot scale. However, they are more difficult to execute, due to the spatial-temporal variation of the sediment sources, as well as the complex interaction between sediment sources, temporary and longer-term sediment sinks and alluvial channels (García-Ruiz 2010; Owens 2005). Scaling effects existed when data of the plot scale were extended to the catchment scale (Fang et al. 2008). Up to the present, few retrospective assessments have been made to measure the amount of the retained runoff and soil by the conservation practices for decades at the catchment scale, especially the soil nutrients. Meanwhile, the data collected from the field were only for short-term monitoring within several years (Minella et al. 2009).

A tremendous difficulty in quantitatively calculating the effects of soil nutrients lies in the fact that little information relates to soil nutrient budgets with altered land-use and climate change settings. Other than soil nutrients, sediment yield was monitored in most of the existing studies, as a parameter of assessing changes in erosion and sediment mobilization resulted from the implementation of soil conservation practices (e.g., Li et al. 2010; Mohammad and Adam 2010; Shen et al. 2010; Minella et al. 2009; Tian et al. 2008; Ran et al. 2004, 2002). Generally, the mean annual soil nutrients retained on various lands were detected to evaluate the effects of alternate land uses (e.g., Jiao et al. 2010; Tripathi and Singh 2009). Quite consistent information had been obtained on the effect of tillage practice and crop choice with the plot studies, whereas that on the route of sediment transport and amount of sediment and nutrients was still scarce, especially that could reach a watercourse from the adjacent slopes (Veihe et al. 2003). Among all the researches, the quality of groundwater in the catchment may be the most direct and convictive parameter. Unfortunately, few results were obtained in the absence of field observation data of nutrients in the groundwater regarding the frequency of occurrence of different erosion processes. Multiplying the rate of nutrient loss by the area of soil conservation practice is another effective way to evaluate the amount of retained soil nutrient (Pan et al. 2006). Nevertheless, only the observation data of soil nutrient loss from the forestland, grassland, slope farmland and some other lands were currently recorded (Zhao et al. 2006), while those from the terrace farmland and dam farmland under different rainfall events were not collected.

Herein we perform a series of calculation on hydro-sediment and nutrient for a period of 50 years undertaken in a representative catchment in the Loess Mesa Ravine Region of the Loess Plateau. The objectives of this study are to (1) quantify the effects of conservation management in terms of retaining soil, water, and especially nutrients, and (2) recommend a preferred conservation practice on the Loess Plateau.

To accomplish these goals was envisioned as a guide for planning sustainable land use and management practices in other areas of the Loess Plateau, China.

7.2 Characteristics of the Nanxiaohegou Catchment

The Nanxiaohegou Catchment (107°37′ E, 35°42′ N; Fig. 7.1) is a representative small watershed in the Loess Mesa Ravine Region of the Loess Plateau, with an area of 36.3 km^2. Its elevation ranges from 1050 to 1423 m. The catchment is composed of mesa and gully, and the area ratio of the mesa to the gully is 57:43. A series of conservation practices have been put into operation in the catchment since the 1950s. The main practices include plantation of trees, establishment of pasturelands, and construction of terraces and check dams. The terrace is a leveled surface on the hill to control the slope erosion, while the check dam is built on the gully to retain hyper-concentration flow from the upper reaches of the catchment. As a huge amount of soil is deposited in the reservoir of the check dam, the dam farmland is formed, where crops or trees could be planted.

7.3 Methods and Materials

Rainfall and runoff were monitored at field sites during the treatment period 1954–2004, except those in 1970, 1986, 1990, 1995–2002. Since 1954, 17 precipitation observation stations, 8 runoff observation stations, and more than 128 runoff plots have been continuously built in the Nanxiaohegou Catchment. Figure 7.2 shows stations and plots currently still in use (Tian et al. 2008). Runoff observation stations were built at the outlets of the gullies to determine the flow rates of runoff and sediment. The amount of erosion from the catchment was measured at the Shibamutai Reservoir runoff observation station near the outlet of the catchment. The inflow and outflow of the reservoir were observed, and the data during the period 1954–2004 were used to assess the conservation practices. As a contrast, a runoff observation station was built at the outlet of the Dongzhuanggou Gully where no control practice was enforced. This gully represents the original state of the catchment before control, where grazing and farming were free, and the vegetation degree of coverage was near 40% in flood season (XFES 1982). In addition, various runoff plots were established in the catchment to explore the erosion processes and evaluate the impact factors. Normal runoff plot was 5 m wide and 16 or 20 m long, while large scale plots usually covered an area above 77,100 m^2. The plots were designed on the basis of various kinds of factors influencing soil and water conservation (Tian et al. 2008). For the woodland, the factors included the coverage, thickness of the litter layer, and operating condition of the structure practice, whereas for the grassland, the factors included the coverage, seed, age, and growth season of the grass. As for the structure measures, e.g., terrace farmland and check dam farmland, the magnitude of retained

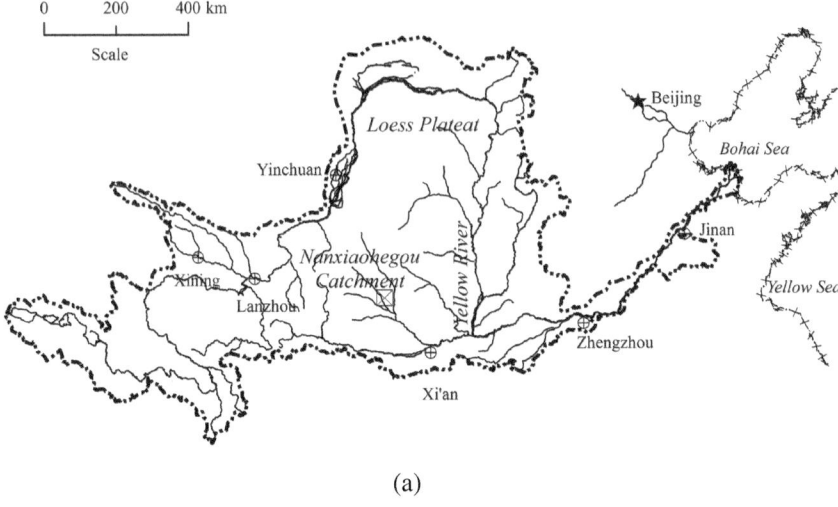

(a)

(b)

Fig. 7.1 Study area. **a** A sketch map of the Loess Plateau of China; **b** A bird's-eye view of the Nanxiaohegou Catchment

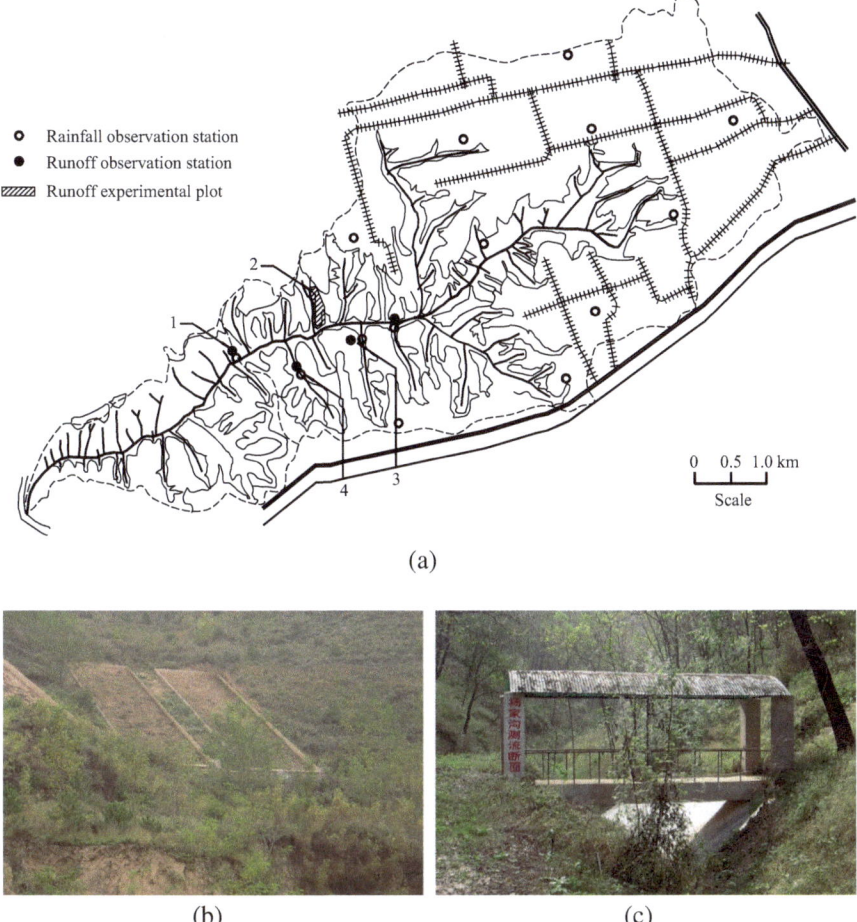

Fig. 7.2 Precipitation observation stations, runoff observation stations, and runoff plots in the Nanxiaohegou Catchment. Adapted from Tian et al. (2008). **a** A sketch map of the Nanxiaohegou Catchment; **b** Runoff plots; **c** Runoff observation stations. Keys: 1. Shibamutai Reservoir runoff & rainfall observation station; 2. Qiugou runoff experimental plots; 3. Huaguoshan Reservoir runoff & rainfall observation station; 4. Dongzhuanggou runoff and rainfall observation station

soil or water was generally calculated according to the reservoir capacity or the farmland area.

In general, the effects of soil and water conservation practices are discussed based on the amounts of retained soil, water, and soil nutrients. In the present work, three indexes are defined to describe the effects: (1) retention ability, representing the amount retained by a certain conservation practice per unit area per year, (2) total retention efficiency (TRE), representing the ratio of the total amount retained by all conservation practices to the total transport amount in a year, and (3) retention ratio, representing the ratio of the amount retained by a certain practice to the total amount retained by all practices in the catchment in a year.

The amount of soil or water aroused by rainfall in the catchment before the conservation practices were built, namely the transport amount of soil or water, includes two parts: (1) the amount of soil or water lost from the catchment, which was determined according to the flow rate and water sampler of the runoff at the catchment outlet, and (2) the amount of soil or water retained by the conservation practices:

$$W = R_a \times A \tag{7.1}$$

where W is the amount of water (m^3/a) or soil ($10^3\ m^3/a$) retained by a certain practice per year; R_a is the retention ability for water ($m^3/(km^2\ a)$) or soil ($10^3\ kg/(km^2\ a)$) of the practice in the catchment; and A is the surface area of the practice (km^2). The area of each practice is based on the statistical yearbook of the catchment, and it is also corrected in accord with other data, e.g., the annual plan or summing-up reports, remote sensing data, data of land use investigation, data of general investigation on soil and water conservation measures, and planning maps.

Since the catchment area is as large as 36.3 km^2, a regional disparity exists as rainfall and runoff are discovered in different rainfall observation stations. Thus, the retention ability of a type of practice in the catchment should be corrected as it is transferred from the test data at a plot scale:

$$R_a = R_{a1} \cdot \alpha \cdot X \tag{7.2}$$

where R_{a1} is the retention ability for water ($m^3/(km^2\ a)$) or soil ($10^3\ kg/(km^2\ a)$) of a certain practice investigated in the plot test; X is the rainfall correction factor for different catchments, and $X = 1$ since the runoff plots were locally set up in the Nanxiaohegou Catchment; α is the rainfall correction factor for different locations in the catchment, $\alpha = P_a/P_{a1}$, where P_a is the precipitation monitored at the test plot, and P_{a1} is the average precipitation in the catchment. In this study, the amount of soil or water retained by a type of practice is quoted from the experimental results reported in the book chapter by Tian et al. (2008), and then the retention ability for water or soil of a certain practice is inversely computed according to Eq. (7.1) on the basis of the amount of retained soil and the area of the practice.

The amount of transported or retained soil nutrients on the land of each conservation practice could be calculated by multiplying the amount of soil retained by the practice with the statistic average nutrient content of the land, so that

$$W_n = W_s \times \overline{C_n}/1000 \tag{7.3}$$

where W_n is the amount of transported or retained total nitrogen (TN), total phosphorus (TP) or SOM per year, 10^3 kg/a; W_s is the amount of transported or retained soil per year, 10^3 kg/a; and $\overline{C_n}$ is the statistic average content of TN, TP or SOM in the soil of the land, g/kg. Since the Loess Plateau is largely in arid or semiarid area, the flow rate of the perennial river in the representative small watersheds, e.g., the Nanxiaohegou Catchment, is so small as to be neglected. As a consequence, the runoff together with the soil nutrients retained by the conservation practice during the flood period is locally absorbed. Thus the amount of nutrient in the soil retained on the dry conservation practice land after an event of representative rainfall is close to that retained during the rainfall. For the Nanxiaohegou Catchment, almost all transported soil has been retained by the following four practices: the dam farmland, terrace farmland, forest land, and grassland. Then $\overline{C_n}$ could be calculated as follows:

$$\overline{C_n} = \left(\sum C_n \cdot W_{Rsi} \right) / \sum W_{Rsi} \qquad (7.4)$$

where C_n is the content of TN, TP or SOM in a certain conservation land, g/kg, and it has been obtained from the representative conservation practices via field sampling and statistic analysis (Zheng et al. 2006). W_{Rsi} is the amount of soil retained by a certain conservation practice, e.g., dam farmland, terrace farmland, forest land and the grassland, 10^3 kg/a.

7.4 Results and Discussion

7.4.1 Distributions of Water, Soil, Nutrients, and Total Retention Efficiency

Distributions of water, soil, and nutrients, and total retention efficiency in the Nanxiaohegou Catchment are shown in Table 7.1. During the treatment period 1954–2004, the average annual amount of runoff loss was 450.9×10^3 m^3, which was markedly less than the average annual transport amount of 961.5×10^3 m^3. Namely, about 53.1% of the transported runoff had been retained. The average annual amount of soil loss was only 14.0×10^6 kg, accounting for 5.2% of the average annual sediment transport amount, and was negligible in comparison with the average annual sediment transport amount. The average TREs for soil all exceeded 90%. At the same time, most soil nutrients had also been retained. The average annual amounts of retained TN, TP and SOM were 78.9×10^3, 326.3×10^3 and 1341.8×10^3 kg, respectively. The average TREs for nutrients were up to 94.8%. What should be noted is that the TRE for water was much smaller than that for soil in the same year.

7.4.2 Retention Ability and Retention Ratio

Contributions of each conservation practice on retaining water, soil and soil nutrients are demonstrated in Table 7.2. Significant differences in the retention effects existed

Table 7.1 Distributions of water, soil, nutrients, and TRE in the Nanxiaohegou Catchment during the treatment period 1954–2004

Year		1954–1959	1960–1969	1970–1979	1980–1989	1990–1999	2000–2004	1954–2004
Precipitation (mm/a)		439.8	409.4	381.2	419.1	349.8	473.9	407.1
Transport amount	Water (10^3 m³/a)	2033.0	852.2	836.5	1394.1	2410.4	1084.7	961.5
	Soil (10^6 kg/a)	94.4	260.1	312.5	332.8	271.7	300.7	268.4
	TN (10^3 kg/a)	29.3	80.9	96.9	103.0	84.1	93.0	83.2
	TP (10^3 kg/a)	121.3	333.9	400.7	426.4	347.9	385.7	344.2
	SOM (10^3 kg/a)	496.1	1375.0	1650.0	1752.1	1434.1	1576.8	1415.4
Precipitation (mm/a)		439.8	409.4	381.2	419.1	349.8	473.9	407.1
Retention amount	Water (10^3 m³/a)	286.7	562.4	543.7	543.7	489.3	658.4	510.5
	Soil (10^6 kg/a)	92.2	253.6	304.7	301.8	258.1	297.1	254.4
	TN (10^3 kg/a)	28.6	78.8	94.5	93.4	79.9	91.8	78.9
	TP (10^3 kg/a)	118.5	325.6	390.6	386.8	330.5	381.0	326.3
	SOM (10^3 kg/a)	484.6	1340.7	1608.7	1589.1	1362.4	1557.9	1341.8
Loss amount	Water (10^3 m³/a)	1746.4	289.7	292.8	850.4	1921.1	426.3	450.9
	Soil (10^6 kg/a)	2.2	6.5	7.8	30.9	13.6	3.6	14.0
	TN (10^3 kg/a)	0.7	2.0	2.4	9.6	4.2	1.1	4.3
	TP (10^3 kg/a)	2.8	8.3	10.0	39.7	17.4	4.6	17.9
	SOM (10^3 kg/a)	11.4	34.4	41.2	162.9	71.7	18.9	73.6
TRE	Water (%)	14.1	66.0	65.0	39.0	20.3	60.7	53.1
	Soil (%)	97.7	97.5	97.5	90.7	95.0	98.8	94.8
	TN (%)	97.7	97.5	97.5	90.7	95.0	98.8	94.8
	TP (%)	97.7	97.5	97.5	90.7	95.0	98.8	94.8
	SOM (%)	97.7	97.5	97.5	90.7	95.0	98.8	94.8

Notes The amount of soil or water retained by each practice taken from Tian et al. (2008)

Table 7.2 Retention ratios of the conservation practices during the treatment period 1954–2004

Land use	Year	RR (%)				
		Water	Soil	TN	TP	SOM
Dam farmland	1954–1959	86.9	94.5	94.5	94.1	95.1
	1960–1969	94.5	98.3	98.0	98.0	98.4
	1970–1979	92.8	98.2	98.1	98.0	98.4
	1980–1989	86.8	96.7	96.8	96.6	97.1
	1990–1999	94.8	98.5	98.6	98.5	98.7
	2000–2004	80.4	93.9	94.2	93.7	94.8
	1954–2004	**90.7**	**97.4**	**97.4**	**97.2**	**97.7**
Terrace farmland	1954–1959	0.2	0.1	0.1	0.1	0.1
	1960–1969	0.2	0.1	0.1	0.1	0.1
	1970–1979	0.8	0.2	0.2	0.2	0.2
	1980–1989	0.9	0.2	0.2	0.2	0.2
	1990–1999	0.5	0.1	0.1	0.1	0.1
	2000–2004	2.3	0.7	0.6	0.7	0.6
	1954–2004	**0.7**	**0.2**	**0.2**	**0.2**	**0.2**
Forest land	1954–1959	10.7	4.5	4.0	4.4	3.7
	1960–1969	2.8	0.9	0.8	0.9	0.7
	1970–1979	4.5	1.2	1.1	1.2	1.0
	1980–1989	11.0	2.7	2.4	2.7	2.2
	1990–1999	4.2	1.2	1.1	1.2	1.0
	2000–2004	15.5	4.7	4.2	4.6	3.9
	1954–2004	**6.8**	**1.9**	**1.7**	**1.9**	**1.6**
Grassland	1954–1959	2.3	0.9	1.5	1.4	1.1
	1960–1969	2.4	0.7	1.2	1.1	0.9
	1970–1979	1.9	0.4	0.6	0.6	0.5
	1980-1989	1.4	0.4	0.5	0.5	0.4
	1990–1999	0.5	0.1	0.2	0.2	0.1
	2000–2004	1.9	0.6	1.0	0.9	0.8
	1954–2004	**1.8**	**0.5**	**0.7**	**0.7**	**0.6**

among the four conservation practices. During the treatment period 1954–2004, the average annual retention ratios for water of check dam farmland, terrace farmland, forest land and grassland were 90.7, 0.7, 6.8 and 1.8%, respectively. With regard to the retention ratio for soil, the check dam farmland, terrace farmland, forest land and grassland were 97.4, 0.2, 1.9 and 0.5%, respectively. The retention ratio of certain practice for soil nutrient was close to that for soil. These results evidently indicate that the check dam farmland played a major role in the control of soil and water loss.

To assess the work capacity of each conservation practice, the retention abilities of the structural measures (e.g., check dam farmland and terrace farmland), and those of the vegetative measures (e.g., forest land and grassland) are also compared and discussed here. As shown in Table 7.3, the retention abilities of the representative

Table 7.3 Retention abilities of the conservation practices during the treatment period 1954–2004

Land use	Year	Area (hm^2)	RA				
			Water (m^3/hm^2)	Soil (10^3 kg/hm^2)	TN (10^3 kg/hm^2)	TP (10^3 kg/hm^2)	SOM (10^3 kg/hm^2)
Dam farmland	1954–1959	–	–	–	–	–	–
	1960–1969	5.3	100 309.43	47 030.19	14.58	60.20	248.79
	1970–1979	5.3	95 207.55	56 433.96	17.49	72.24	298.54
	1980–1989	10.7	44 088.79	27 274.77	8.46	34.91	144.28
	1990–1999	10.7	43 370.09	23 766.36	7.37	30.42	125.72
	2000–2004	10.7	49 492.52	26 082.24	8.09	33.39	137.98
	1954–2004	**8.5**	**54 195.55**	**29 008.20**	**8.99**	**37.13**	**153.45**
Terrace farmland	1954–1959	3.2	178.13	31.25	0.01	0.04	0.13
	1960–1969	4.6	263.04	39.13	0.01	0.05	0.17
	1970–1979	24.6	178.86	30.08	0.01	0.04	0.13
	1980–1989	17.2	275.00	43.60	0.01	0.05	0.18
	1990–1999	16.9	130.77	21.30	0.01	0.03	0.09
	2000–2004	31.1	476.85	67.20	0.02	0.08	0.28
	1954–2004	**16.3**	**234.22**	**36.89**	**0.01**	**0.05**	**0.16**
Forest land	1954–1959	170.4	179.23	24.30	0.01	0.03	0.11
	1960–1969	114.7	139.15	19.35	0.01	0.02	0.08
	1970–1979	243.9	100.62	15.09	0.00	0.02	0.07
	1980–1989	379.8	157.48	21.48	0.01	0.03	0.09
	1990–1999	294.4	69.80	10.84	0.00	0.01	0.05
	2000–2004	439.8	231.47	31.81	0.01	0.04	0.14
	1954–2004	**273.8**	**126.68**	**17.75**	**0.00**	**0.02**	**0.08**
Grassland	1954–1959	52.7	123.15	16.51	0.01	0.03	0.10
	1960–1969	116.5	117.00	16.22	0.01	0.03	0.10
	1970–1979	117.0	87.18	10.17	0.00	0.02	0.06
	1980–1989	62.9	117.65	16.85	0.01	0.03	0.10
	1990–1999	50.6	49.41	5.53	0.00	0.01	0.03
	2000–2004	78.2	156.52	24.30	0.01	0.05	0.15
	1954–2004	**79.7**	**115.51**	**15.44**	**0.01**	**0.03**	**0.09**

Notes No observation data for check dams in 1950s, since the dam farmlands had not been formed on a relatively large scale at that time conservation practices were in the following order: check dam farmland > terrace farmland > forest land & grassland. The check dam was the most appropriate practice to conserve soil and water. Its retention abilities for water, soil, TN, TP and SOM were 231–469, 786–1878, 871–1854, 805–1656, and 984–1982 times of other practices, respectively. The retention ability of the forest land was approximately equal to that of the grassland, and the former was a little larger than the later in the Nanxiaohegou Catchment due to the flourishing trees.

7.4.3 Positive Effects of the Conservation Practices

Understanding the complexity of land-use and land-cover changes and their driving forces and impacts on human and environmental security is important for the planning of natural resource management and associated decision making (Garedew et al. 2009). Soil cover and rainfall intensity are considered to have significant impacts on soil erosion, and close relationships exist among them (Jin et al. 2009). We have previously described that four representative conservation practices have been built in the Nanxiaohegou Catchment, namely, check dam farmland, terrace farmland, forest land, and grassland. Vegetation cover can increase infiltration, reduce surface runoff, and thus markedly retard sheet erosion (Woo and Luk 1990). The retention ability of the forest land was influenced by many factors, including the slope gradient and length, the tree variety and age, plant cover structure and density, and the same case happened to the grassland. Terracing and contour ridges caused an increase in infiltration rate, largely because water would need more time to infiltrate into the ground (Al-Seekh and Mohammad 2009). Thus, terrace farmland can also effectively relieve soil and water loss from the slope. Different to the measures controlling slope erosion above-mentioned, some practices, e.g., check dams and gully head protections, could control soil and water loss both from the slope and the gully. The check-dam can slow down the velocity of the sediment-laden water, resulting in a lower sediment transport capacity, and therefore some of the sediments could sink down (Xu et al. 2006). The main factors affecting the retention ability of the check dam are supposed to be the dam height and the reservoir capacity.

Data presented here can be used to explain the commonly observed phenomenon that soil erosion and nutrient losses have been greatly decreased with the implementation of the various conservation practices. In the Nanxiaohegou Catchment during the treatment period 1954–2004, about 510.5×10^3 m^3 of transported water, half of the total, had been retained by the conservation practices every year. Simultaneously, almost all soil and nutrients, 94.8% of the total transported soil and nutrients, had been locally retained in the catchment. As indicated in Table 7.1, soil and water loss was not only markedly controlled, but also kept at a relatively low level. It can be assumed that hydro-sedimentologic balance had been realized in the Nanxiaohegou Catchment, similar to relative stability of the check dam system (Xu et al. 2006). As a result, the qualities of the runoff and ground water had been meliorated, the land had become more fertile, and the eco-environment had been greatly improved.

Table 7.1 also shows that, TREs for soil and water were relatively lower in the 1980s or 1990s. This may be because of the changes in the conservation practices and precipitation, especially the former, since then no other obvious nonstationarity, e.g., urbanization, existed. It has been mentioned above that most of the check dams, just as other conservation practices, were built in the 1960s–1970s. They were gradually filled up in the following 10 years, and consequently retained less soil and water from then on. However, in recent years, some new reservoirs and check dams have been built, and the TREs grow up therefrom.

On the other hand, enormous amounts of soil retained from small watersheds by the conservation practices do not only form large-scale fertile farmland but also safeguard the Yellow River against overflow. The Yellow River (as shown in Fig. 7.1), with a total length of 5464 km, is the second longest river in China. Due to serious silting, levees along the lower reaches have to be heightened every year, thus an Above Ground River is formed with a riverbed 10 m higher than the surrounding ground. In fact, most sediment in the lower Yellow River comes from the Loess Plateau. Hence, reducing sediment pouring into the Yellow River from the Loess Plateau is the most fundamental solution to safeguard the river (Xu et al. 2004). It has been revealed in the above calculation that, in the Nanxiaohegou Catchment, about 254.4×10^3 kg of soil per year, and a total of $10\,176.8 \times 10^3$ kg of soil during the period 1954–2004, had been retained by the conservation practices, which would be a huge threat if all of them were joined into the river. Thus, we may reasonably come to the conclusion that a big contribution has been made in relieving sediment accumulation in the Yellow River by the conservation practices on the Loess Plateau.

7.4.4 Negative Effects of the Conservation Practices

Soil and water conservation practices might negatively influence the water cycle process in the catchment through the following ways: (1) to bring impact on the ecological environment and probably aggravate the problem o f bcal water resources in the catchment, and (2) to decrease runoff and even cut-off the Yellow River.

Some scientists, e.g., Chen et al. (2007), insisted that the long-term ecological effects of soil and water conservation on the Loess Plateau were still not clear. In light of the statistics, the survival rate of trees was only about 25% on the Loess Plateau (Li 1997), and the growth quality of survival trees was also not so good due to the poor survival conditions (Xu et al. 2004). The evapotranspiration of perennial forest and grass vegetation was greater than or equal to the annual precipitation on the Loess Plateau. Thus water deficit had to be compensated by absorbing water in deep soil, which might lead to the formation of the dried soil layer, and further the low survival rates of forest and grass (Li 1997). Nevertheless, soil loss hazards might be alleviated if combined conservation measures, including terraces and contour tillage, were applied in the watershed (Shi et al. 2004). Anyhow, as a widespread structure practice, the check-dam system in gullies is one of the most effective ways to conserve soil and water on the Loess Plateau.

Jing and Zheng (2004) and Li (1997) have reported that surface runoff retained by the structure measures of soil and water conservation, was often consumed by local humans and animals in the small watersheds of the upper reaches, which would result in a decrease of the total runoff, and even lead to a cutoff of the lower Yellow River. However, in this research, an assessment of regional soil and water loss has indicated the successfulness of the conservation efforts by quantifying the contribution of improved land cover to potential erosion reduction. The notable difference has been obtained in the TRE for water and that for soil in the same practice. It has been shown in Table 7.1 that the former was much smaller than the later. The result of relatively small TRE for water in this study testified that the conservation practices worked better in retaining sediment than in reducing runoff from the Loess Plateau. On the other hand, because of the accumulation of silt in the Yellow River channel, water has to be used for flushing out the sediments deposited on the channel bed and transporting them to where it needed or into the Bohai Sea (Yu 2006). Thus water for sluicing the soil could be put into other uses in the lower reaches of the Yellow River in virtue of the conservation practices on the Loess Plateau.

From what has been discussed above, we may safely draw the conclusion that soil and water conservation practices will not aggravate the shortage of water resources, either in the local catchment or in the Yellow River. Upon the present condition, negative effects of the conservation practice are not noteworthy.

7.4.5 Role of Check Dam in Reducing Soil and Water Loss

Soil conservation measures should be selected referring to many factors, such as soil types, topographical conditions, hydrological processes, land uses and land management practices (Shen et al. 2010). In accordance with the specific circumstance of the Loess Plateau, the emphasis of conservation may be slope control or gully control, and planting or structure practice, but which one is more effective in preventing the sediment poured into the Yellow River from this area, is still in dispute. In semiarid Tunisia, more emphases have been imposed on terracing than gully control (Bouchnak et al. 2009). While in the Three Gorges Reservoir Area of southern China, hedgerows were the best way for soil erosion control, followed by stone dike terraces and soil dike terraces (Shen et al. 2010). Yet, in any case, the suitable vegetative cover was a primary consideration for soil and water conservation (Mohammad and Adam 2010). For instance, in the Upper Min River watershed of Southwest China, vegetation played a crucial role in protecting soil from erosion loss (Zhou et al. 2008), and in the Ziwuling region of the Loess Plateau, vegetation was regarded as a key factor affecting soil erosion (Zheng et al. 2006).

However, most regions of the Loess Plateau are short of water and thus unsuitable for planting. Control of soil and water erosion on the Loess Plateau by planting alone has been proved unsuccessful in recent decades (Xu et al. 2004). Trees are difficult to survive owing to the arid climate and barren soil, and even if they are alive, most of them will not grow strong enough to control soil and water losses. As revealed in this

Fig. 7.3 A check dam and a dam farmland in the Nanxiaohegou Catchment

study, in the Loess Mesa Ravine Region of the Loess Plateau, check dams (Fig. 7.3) had the best effect on controlling soil erosion, followed by the terrace farmlands, and then the forest lands and the grasslands. More than 90% of the retained soil and water, especially soil, was accomplished with the dam farmland, although it occupied only 2.3% of the total area of all conservation practices. Similar results were also obtained by other researchers. For example, it has been reported that at the global scale, sediment loads are chiefly reduced thanks to sediment trapping by dams (e.g. Minella et al. 2009; Syvitski et al. 2005; Walling 2005). Xu et al. (2004) has also pointed out that the check dam system in gullies was the most effective way to conserve soil and water on the Loess Plateau. On the whole, as a representative gully control practice of conserving soil, water, and nutrients, the check dam is much more preferred.

7.4.6 *Future Research Scenarios*

Toward future planning, more studies are still needed for this subject, including: (1) Observation of water quality of the groundwater in the selected catchment and the runoff through the outlet of the catchment. Other than the amount of retained nutrients, the nutrient contents of the groundwater and the runoff are the most direct indicators reflecting the effects of conservation practices on water environments. Nevertheless, few observation data on water quality have been obtained until the present. (2) Application of the distributed hydrologic models. The structures of distributed hydrologic models have become very mature, and the models have been broadly applied in China. Yet, soil erosion is still very serious due to the broken topography and intense rainfall on the Loess Plateau. Thus, it is urgent to choose a suitable distributed hydrological model, and then develop it according to the characteristics of the region, in order to make the simulation results agree with the prototype phenomenon as far as possible. (3) More efforts to develop management systems favoring soil conservation and maintaining economic viability. It is worth planting various cash crops on the dam farmland, for the farmland formed by check dams is very suitable for planting in virtue of the fertile soil and abundant water.

7.5 Conclusions

This chapter presents the results of a catchment scale investigation on hydro-sediment and nutrient for a period of 50 years undertaken in a representative catchment in the Loess Mesa Ravine Region of the Loess Plateau, China, where various conservation practices were constructed, leading to notable changes in soil and nutrient yield. Soil erosion and nutrient losses had been greatly decreased, testifying that the conservation practices were effective in conserving the limited water resources and reducing soil erosion which threatens the sustainable development in this area. The check dam play a major role in controlling soil and water loss, for relatively small variations in the area of check dam farmland can influence erosion behavior profoundly. On the other hand, the TRE for water was much smaller than that for soil, indicating that the conservation practices worked better in retaining sediment than in reducing runoff on the Loess Plateau. Hence the negative effects of the conservation practices on reducing water to the Yellow River were relatively slight.

References

Al-Seekh S H, Mohammad A G. 2009. The effect of water harvesting techniques on runoff, sedimentation, and soil properties. Environmental Management, 44(1): 37–45.

Bouchnak H, Felfoul M S, Boussema M R, et al. 2009. Slope and rainfall effects on the volume of sediment yield by gully erosion in the Souar lithologic formation (Tunisia). Catena, 78(2): 170–177.

Castillo V M, Martinez-Mena M, Albaladejo J. 1997. Runoff and soil loss response to vegetation removal in a semiarid environment. Soil Science Society of America Journal, 61(4): 1116–1121.

Chen L D, Wei W, Fu B J, et al. 2007. Soil and water conservation on the Loess Plateau in China: review and perspective. Progress in Physical Geography, 31(4): 389–403.

Fang H Y, Cai Q G, Chen H, et al. 2008. Effect of rainfall regime and slope on runoff in a gullied loess region on the Loess Plateau in China. Environmental Management, 42(3): 402–411.

Fu B J, Meng Q H, Qiu Y, et al. 2004. Effects of land use on soil erosion and nitrogen loss in the hilly area of the Loess Plateau, China. Land Degradation and Development, 15(1): 87–96.

García-Ruiz J M. 2010. The effects of land uses on soil erosion in Spain: A review. Catena, 81(1): 1–11.

Garedew E, Sandewall M, Söderberg U, et al. 2009. Land-use and land-cover dynamics in the central rift valley of Ethiopia. Environmental Management, 44(4): 683–694.

Jiao J G, Yang L Z, Xu J X, et al. 2010. Land use and soil organic carbon in China's village landscapes. Pedosphere, 20(1): 1–14.

Jin K, Cornelis W M, Gabriels D, et al. 2009. Residue cover and rainfall intensity effects on runoff soil organic carbon losses. Catena, 78(1): 81–86.

Jing K, Zheng F L. 2004. Effects of soil and water conservation on surface water resource on the Loess Plateau. Research of Soil and Water Conservation, 11(4): 11–12,73 (in Chinese).

Li Y S. 1997. Relation between control in Loess Plateau and no-flow in the Yellow River. Bulletin of Soil and Water Conservation, 17(6): 41–45 (in Chinese).

Li C B, Qi J G, Feng Z D, et al. 2010. Quantifying the effect of ecological restoration on soil erosion in China's Loess Plateau region: an application of the MMF approach. Environmental Management, 45(3): 476–487.

Maass J M, Jordan C F, Sarukhan J. 1988. Soil erosion and nutrient losses in seasonal tropical agroecosystems under various management techniques. Journal of Applied Ecology, 25(2): 595–607.

Meng Q H, Fu B J, Yang L Z. 2001. Effects of land use on soil erosion and nutrient loss in the Three Gorges Reservoir Area, China. Soil Use and Management, 17(4): 288–291.

Mills W C, Thomas A W, Langdale G W. 1992. Seasonal and crop effects on soil loss and rainfall retention probabilities: an example from the U.S. Southern Piedmont. Soil Technology, 5(1): 67–79.

Minella J P G, Merten G H, Walling D E, et al. 2009. Changing sediment yield as an indicator of improved soil management practices in southern Brazil. Catena, 79(3): 228–236.

Mohammad A G, Adam M A. 2010. The impact of vegetative cover type on runoff and soil erosion under different land uses. Catena, 81(2): 97–103.

Owens P. 2005. Conceptual models and budgets for sediment management at the river basin scale. Journal of Soils and Sediments, 5(4): 201–212.

Pan Y J, Yang A M, Wang Y J, et al. 2006. Effects of soil and water conservation on controlling non-point pollution in Yanan, Shannxi. Technology of Soil and Water Conservation, (6): 34–36 (in Chinese).

Quinton J N, Edwards G M, Morgan R P C. 1997. The influence of vegetation species and plant properties on runoff and soil erosion: results from a rainfall simulation study in south east Spain. Soil Use and Management, 13(3): 143–148.

Ran D C, Liu B, Wang H. 2002. Analysis on sediment reduction of the Yellow River through soil and water conservation measures. Soil and Water Conservation in China, (10): 35–37 (in Chinese).

Ran D C, Luo Q H, Liu B, et al. 2004. Effect of soil-retaining dams on flood and sediment reduction in middle reaches of yellow river. Journal of Hydraulic Engineering, 35(5): 7–13 (in Chinese).

Shen Z Y, Gong Y W, Li Y H, et al. 2010. Analysis and modeling of soil conservation measures in the Three Gorges Reservoir Area in China. Catena, 81(2): 104–112.

Shi Z H, Cai C F, Ding S W, et al. 2004. Soil conservation planning at the small watershed level using RUSLE with GIS: a case study in the Three Gorge Area of China. Catena, 55(1): 33–48.

Syvitski J P M, Vörösmarty C J, Kettner A J, et al. 2005. Impact of humans on the flux of terrestrial sediment to the global coastal oceans. Science, 308(5720): 376–380.

Thornton K W, Holbrook S P, Stolte K L, et al. 2000. Effects of forest management practices on mid-Atlantic streams. Environmental Monitoring and Assessment, 63(1): 31–41.

Tian X F, Jia Z X, Liu B, et al. 2008. An analysis on the benefit of soil and water conservation and rule of soil and water loss in the representative catchment of the Loess Mesa Ravine Region. Zhengzhou: Yellow River Conservancy Press (in Chinese).

Tripathi N, Singh R S. 2009. Influence of different land uses on soil nitrogen transformations after conversion from an Indian dry tropical forest. Catena, 77(3): 216–223.

Veihe A, Hasholt B, Schiøtz I G. 2003. Soil erosion in Denmark: processes and politics. Environmental Science and Policy, 6(1): 37–50.

Walling D E. 2005. Tracing suspended sediment sources in catchments and river systems. Science of the Total Environment, 344(1–3): 159–184.

Woo M K, Luk S H. 1990. Vegetation effects on soil and water losses on weathered granitic hillslopes, South China. Physical Geography, 11(1): 1–16.

XFES (Xifeng Soil and Water Conservation Experimental Station of the Yellow River Conservancy Commission). 1982. Experiments on soil and water conservation (unpublished research reports): 145–148 (in Chinese).

Xu X Z, Zhang H W, Zhang O Y. 2004. Development of check-dam systems in gullies on the Loess Plateau, China. Environmental Science and Policy, 7(2): 79–86.

Xu X Z, Zhang H W, Wang G Q, et al. 2006. A laboratory study on the relative stability of the check-dam system in the Loess Plateau, China. Land Degradation and Development, 17(6): 629–644.

Yu L S. 2006. The Huanghe (Yellow) River: recent changes and its countermeasures. Continental Shelf Research, 26(17): 2281–2298.

Zhao H B, Liu G B, Cao Q Y, et al. 2006. Influence of different land use types on soil erosion and nutrition care effect in loess hilly region. Journal of Soil and Water Conservation, 20(1): 20–24,54 (in Chinese).

Zheng B M, Tian Y H, Wang Y, et al. 2006. Construction theory and practice of soil and water conservation ecological construction project in the Jiuyuangou demonstration area of the Yellow River Basin, Zhengzhou: Yellow River Conservancy Press (in Chinese).

Zhou P, Luukkanen O, Tokola T, et al. 2008. Effect of vegetation cover on soil erosion in a mountainous watershed. Catena, 75(3): 319–325.

Chapter 8
Sediment-Storage Effects of Check-Dam System in the Small Watershed

Abstract The building of check-dams is one of the most effective measures for the conservation of soil and water on the Loess Plateau of China, and the hydro-sedimentologic balance is the most important factor influencing the relative stability of the check-dam systems. This means that soil and water in the small watersheds controlled by the check-dams will be absorbed internally, without the need of raising the height of the dams, if some given parameters have reached certain values. A runoff simulation experiment for a single check-dam and a rainfall simulation experiment for the whole check-dam system had been conducted to simulate the induced morphological changes affecting the stability of check dam systems. The results indicate that the main reasons causing the check-dam to show good relative stability are the enlargement of the dam-land area, the alleviation of erosion by the check-dam, and the auto-stabilizing mechanism of the gullies.

Keywords Loess Plateau · Check-dam · Relative stability · Model experiment · Simulation

8.1 Relative Stability and Optimum Programming of Check Dams on the Loess Plateau

The Loess Mesa Ravine Region and the Loess Hill Ravine Region, covering 200,000 km² of the Loess Plateau in China, are areas with the highest erosion rates on earth. The best method of conserving soil and water in these areas is to construct check dam systems in gullies. Currently, more than 100,000 check dams have been built in these regions and the amount of sediments retained by check dams is the largest among all measures (Xu et al. 2004a). Moreover, the Chinese Ministry of Water Resources is planning to construct an additional 163,300 check dams in this area before 2020 (MWRC 2003). Establishing an optimal design and sequence to construct check dams would accelerate the construction of check dam systems, reduce the engineering costs, and enhance sediment retention in the local watershed (Tian et al. 2003; Wu and Huang 1995). Such an optimal design would generate the number, location and capacity of check dams for a watershed, and the erosion rate in the controlled area.

© Science Press and Springer Nature Singapore Pte Ltd. 2020
X. Xu et al., *Experimental Erosion*,
https://doi.org/10.1007/978-981-15-3801-8_8

Designing check dam system requires an estimate of (1) preferred dam sites; (2) number of dams required and their heights for sediment interception and flood detention; (3) the optimal sequence and interval for dam construction (Wu 1994). Mathematical models have recently been utilized for designing check dam systems (e.g., Wu and Huang 1995; Wan et al. 1995; Lin et al. 1995). However, several limitations exist in these models. For example, the Linear Programming Method (Wu 1994) is a static model that produced results biased for the established condition whereas the small watershed is an open and dynamic system. Although it is dynamic and multi-objective, the Nonlinear Programming Method, another mathematical model, is not credible enough to be applied in practical engineering (Li et al. 1995). To minimize the number of decision-making variables, only skeleton dams can be analyzed using current mathematical models (Wu 1994).

Relatively stability of the check dam system plays an important role in the dam design. The fact that the check-dam systems on the Loess Plateau of China always exhibit a good relative stability is a unique geomorphological phenomenon. It has been found that the Loess Mesa Ravine Region and the Loess Hill Ravine Region, which cover 200,000 km^2 of the Loess Plateau in China, are among the areas having the highest erosion rates on earth, and the most effective way to conserve soil and water in these areas is to build check-dam systems in the gullies. At present, more than 100,000 check-dams have been built, and the amount of sediments retained by the check-dams is the largest among all measures (Xu et al. 2004a). Moreover, official documents from the Chinese Ministry of Water Resources have stated that 163,300 check-dams will be built in this area before 2020 (MWRC 2003). Farmers would prefer to plant crops in the deposited areas behind the dams, which is called the dam-land, owing to their more abundant water content and more fertile soil as compared with the slope land. According to field investigations on the check-dam systems of the Shanxi Province (Fang 1996), 6000–7500 kg grains could be harvested per hectare of the dam-land, which was 8 to 10 times higher than those on the slope land. At present, the dam-land occupies only 9% of the whole farmland areas in the Loess Hill Ravine Region, but their grain yields comprise 23.5% or higher of the over-all grain production of this region (Xu and Wang 2000). However, when tens or hundreds of check-dams are built in a small watershed area not larger than 100 km^2, it will present quite complicated hydro-sedimentologic problems. Elements such as the layout of the dam-sites, the heights of the dams and the areas controlled by the small watersheds should all meet the requirements of relative stability during the design and construction of the check-dam systems. The concept of "relative stability of the check-dam system" first emerged in the 1960s, which was stemming from the recognition of the key factors that determined the natural balance of the check-dams. Inspired by the natural Juqiu, a kind of dam-land resulting from land-slips which never overflows when impounding the floodwater and soil for centuries, people came to recognize the important role of dam height and the ratio of dam-land area to that of the controlled watershed. It was observed that if the above parameters have reached certain values, the soil and water in the small watershed could be internally absorbed, without the need of raising the height of the dam. Fang (1995) has suggested that check-dam systems exhibiting good relative stability have to fulfill the following requirements: (1) No flooding will occur. That is, it should be ensured that the check-dam system can withstand rainstorms. (2) Guarantee of harvest from the dam-land, which implies the reducing of losses of the planted crops due to rainstorms. (3) Conservation of floodwater and sediment via impounding. (4) Increasing of the dam height and repairing of the dams after prolonged utilization are unnecessary. To attain relative stability of the check-dam system to meet these purposes, many elements have to be considered: hydrological, geographical and geological conditions of the controlled small watersheds, area of the dam farmland, varieties of the crops, etc. Empirically, the check-dam becomes relatively stable when the ratio of the dam farmland area to that of the controlled watershed is between 1:25 and 1:15. When the impounded water depth is less than 0.8 m and the storage time is shorter than 3–7 days, a dam designed for enduring a 100-year storm in a century is found to be relatively stable (Zeng et al. 1995).

The coefficient for relative stability of the dam system I is defined as the ratio of the area of dam farmland to that of the watershed corresponding to flood frequency f. When the flood frequency f is 2% and the depth of impounded water in the dam farmland is equal to the critical value, the coefficient is called the critical coefficient for relative stability of the dam system I_C. If $I > I_C$, the dam system is stable, and if $I < I_C$, the dam system is not stable.

$$I_C = \frac{W_p}{\delta \cdot F} \tag{8.1}$$

where W_p is the amount of impounded floodwater (m^3) when the flood frequency f is 2%; δ is the design depth of water for the dam farmland (m); and F is the area of controlled watershed by the check-dam (km^2) (Zeng et al. 1999).

Zeng et al. (1995) has confirmed the feasibility of maintaining the relative stability of the check-dams by comparing the relationship between the dam height and the retention area in the Wangjiagou watershed of China. Fang et al. (1998), Fang (1995) and Zeng et al. (1999) have studied the condition, criterion and mechanism for the maintaining of the relative stability by investigating hundreds of typical check-dams on the Loess Plateau. Empirically, in small watersheds of the Loess Plateau, the check-dam systems will remain relatively stable when the ratio of the dam-land area to the area of the controlled watershed is between 1/25 and 1/15. When the impounded water depth is less than 0.8 m and the storage time is shorter than 3–7 days, a dam designed for withstanding a rainstorm occurring once in a hundred year has been found to be relatively stable (Zeng et al. 1995). Lei and Zhu (2002) have developed a mathematical model for optimizing the layout of the dams in a small watershed according to the principle of relative stability. Nevertheless, dissensions still exist on the theory of relative stability of check-dams. Li (2004) expressed the idea that no check-dam would be stable if sediment has to be detained in the reservoir when soil and water is inputted continuously from the upper reaches.

Alternatively, scaled model laboratory experiments may be a useful method that produces results that can be employed to design check dam systems. The use of physical models to test or predict the performance of full-scale prototype behavior has several inherent advantages such as (a) to give insight into problems which cannot be solved theoretically or numerically; (b) to get better control of boundary conditions; and (c) to save time and money (Timmons 1984). The accuracy of such predictions depends on the similarity between model and prototype in three principal characteristics: (a) geometric similitude; (b) kinematic similitude; and (c) dynamic similitude. While some similarity requirements appear to have been firmly established, others have not been fixed (Zhang et al. 1994; Timmons 1984).

Scaled modeling has been widely used for a long time in the field of hydraulics and river engineering (e.g., Zhang et al. 1994). However, few studies simulated the process of soil loss using a scaled model experiment, as such simulation is quite complicated in that rainfall, soil surface crusting, land-use and vegetation cover, etc., should be considered. Hancock and Willgoose (2004) investigated the effect of erosion on a back-filled and capped earthen dam wall by constructing an experimental model landscape simulator in a laboratory. The design of the rainfall simulator was difficult to directly scale the rainfall runoff processes to the field conditions. Consequently, no attempt was made to match the rate of gully development on the tailings dam to field-scale processes. Scaled model experiments for soil and water erosion in small watersheds have recently been conducted and progress has been made in applying similitude methodology. Shi et al. (1997a, b) observed the quantitative erosion in gullies and on slopes in a small watershed model of the landform of Xiaofanjiagou Gully in Shaanxi Province (the length scale was 75). Jiang et al. (1994) and Yuan et al. (2000a, b) who carried out a series of model experiments for different degrees of erosion control in a small watershed on the Loess Plateau investigated the relation between runoff and sedimentation. Rainfall and soils similar to those of the prototype were applied, and the landforms were scaled. Nevertheless, none of these studies used an explicit scale relation between the prototype and the model.

Xu and Zhang (2004) presented a new method to simulate soil and water processes in small watersheds of the Loess Plateau, and has tested the method using serial model experiments. For this method, the ratio of model geomorphological evolvement rate to the corresponding prototype evolvement rate is kept constant so that soil and water erosion processes of the prototype can be reflected by model experimental results. In their work, the erosion ratio between the model and prototype-R is given by Eq. (8.2):

$$\frac{\overline{H_{E_{m(1)}}}}{L_{m(1)}} \bigg/ \frac{\overline{H_{E_{p(1)}}}}{L_{p(1)}} = \frac{\overline{H_{E_{m(2)}}}}{L_{m(2)}} \bigg/ \frac{\overline{H_{E_{p(2)}}}}{L_{p(2)}} \cdots \frac{\overline{H_{E_{m(i)}}}}{L_{m(i)}} \bigg/ \frac{\overline{H_{E_{p(i)}}}}{L_{p(i)}} = R(\text{Const}) \qquad (8.2)$$

where i is the run of rainfall; $\overline{H_E}$ is the mean score/deposition elevation of the land cover (m); L is the length of the small watershed (m); the subscript p and m represent the prototype and model, respectively. $\overline{H_E}/L$ is a dimensionless parameter for the accumulated erosion after a single rainfall. R is the ratio of erosion between the prototype and the model. For experiments with prototype soil, the volume scale is shown as follows:

$$\lambda_S = \lambda_L^3 / R \qquad (8.3)$$

where λ_S is the volume scale and λ_L is the length scale. Equations (8.2) and (8.3) illustrate that erosion by a single simulated rainfall in the scaled model experiment is equal to that in the prototype when $R = 1$. When $R > 1$, the model sediment volume should be multiplied by a smaller coefficient to estimate sediment volume for the prototype; when $R < 1$, the model sediment volume should be multiplied by a larger coefficient to estimate sediment volume for the prototype. Compared with sediment transport capacity and sedimentation similitude in river engineering, the conditions of kinematic similitude and dynamic similitude are embodied in the scale of cumulative sediment volume for each rainfall event (Eq. 8.3). In this study four measures are recommended to ensure that the ratio R remains constant: (1) A bare land model with a corresponding erosion rate is applied in the experiment. (2) Dimensions of the landform, including the check-dam, are normally scaled down according to the prototype watershed. (3) A soil similar to that in the prototype is used. (4) Antecedent water content before each rainfall simulation is constant.

8.2 Characteristics of the Yangdaogou Catchment

As the prototype catchment or the runoff experiments, the Yangdaogou Catchment (Fig. 8.1) is a typical small watershed located in the Loess Hill Ravine Region covering an area of 0.206 km^2. The total length of the gullies is 752 m, and 240 m of which are from the outlet, with a "U"-shaped gradient of 2.3%. The ground surface is

Fig. 8.1 Headwater of the Yangdaogou Catchment

covered with the Lishi loess and the Malan loess (Cai et al. 1998). The precipitation in this area can be found in the general condition of the Loess Plateau mentioned above. In the 1960s, no dams had been built in this region, so that the average annual runoff rate was 36,700 m^3/km^2, and the average annual soil loss rate was 20,811 t/km^2 (Zhang et al. 1995). Therefore, annual soil loss, namely, the soil loss of the conceptual prototype in a rainfall event, S_p, could be obtained by the multiplying the watershed area by annual soil erosion rate, which is 4.29×10^6 kg/a. As in engineering practices, the annual erosion rate is adopted as an index when planning the check dam system, and a rainfall event in the recapitulated prototype watershed accounts for soil loss in one year in the prototype watershed.

The prototype catchment of the rainfall experiments is abstracted according to the common erosion characteristics of the Loess Plateau. The conceptual watershed is also located in the Loess Hill Ravine Region, but it covers an area of 3.32 km^2. The total length of the gullies is 3 km, and 694 m of which are from the outlet, with a "U"-shaped gradient of 2.3%. Soil of the ground surface, the average annual runoff rate, and average annual soil loss rate in this small watershed are all correspondingly same to those in the Yangdaogou Catchment. The soil loss of the conceptual prototype catchment in a rainfall event, S_p, is 6.87×10^7 kg/a.

8.3   Experimental Methods

Scaled-down models are scale models of a general class that possess geomorphic features, which can be regarded as the replica of a natural landform, but scaled down in such a way that ratios of significant dimensions and forces are equal to those in nature (Timmons 1984). Three conditions should be satisfied to ensure physical similarity: (a) geometric similitude; (b) kinematic similitude; and (c) dynamic similitude. While some similarity requirements seemed to have been firmly established, others have not been fixed so far (Timmons 1984; Albertson et al. 1960). Scale modeling has been widely used and has a long history in the field of hydraulics and river engineering (Zhang 1994). However, few researchers have published data on the simulating of the process of soil loss by scale model experiments, as it is very complicated with respect to rainfall, soil surface crusting, land-use and vegetated cover, etc. Hancock and Willgoose (2004) have examined the effect of erosion on a back-filled and capped earthen dam wall by constructing an experimental model landscape simulator in the laboratory. However, the designing of a rainfall simulator for directly scaling the processes of rainfall runoff to the fields is rather difficult. Consequently, no attempts have been made to match the rate of gully development on the tailings dam to the field-scale processes. Recently, scale model experiments on soil and water erosion in small watersheds have been performed, and progress has been made in similitude methodologies. Shi et al. (1997a, b) have observed quantitative erosions in gullies and on slopes of a small watershed model shrunk from the landform of the Xiaofanjiagou Gully of the Shaanxi Province in China (the length scale was 75). Jiang et al. (1994) and Yuan et al (2000a, b) have carried out a series of scale-down model experiments for varying extents of erosion control in designated small watersheds of the Loess Plateau to investigate the relationship between the runoff and the sedimentation. Rainfalls and soils similar to those of the prototype were applied, and the landforms were scaled down according to the prototype in the above-mentioned experiments. Nevertheless, no explicit scale relationship between the prototype and the model has been reported in the above studies.

Xu et al. (2009) has proposed a semi-scale physical model experimental method to simulate the soil and water process in small watersheds of the Loess Plateau, and confirmed it with serial model experiments. The key idea of the method is that the ratio of the modeled geomorphological evolvement rate to the corresponding rate of the prototype should be kept constant after the determined runs of the rainfalls, so that the soil and water erosion processes of the prototype could be represented by the result of the model experiment. Four measures have been proposed to ensure the ratio R a constant: (1) A bare land model with a corresponding erosion rate should be applied in the experiments. (2) Dimensions of the landform, including the check-dams, should normally be scaled down according to the prototype watersheds. (3) Soil similar to those of the prototype should be used. (4) The antecedent water content before each rainfall simulation should be kept constant. The proposed method provides the quantitative proportion of soil loss between the prototype and model, using similarity criteria. Nevertheless, it did not strictly meet the similar situation in the conventional scale model experiment and could not provide the scale relationship of the erosion depths between the model and the prototype, etc.

Two experiments were carried out to demonstrate the hydro-sedimentologic balance on the dam-land, one for a single check-dam via runoff simulation, and the other for the check-dam system via rainfall simulation. The objectives of this study are to prove the feasibility of the relative stability idea for check-dams from the point of view of hydro-sedimentologic balance, and present a method for designing scale model experiments on soil and water erosions.

8.3.1 Runoff Simulation

Generally, according to the correlative criterion (MWRC 2009), middle scale check-dams with dam heights of 15–25 m and reservoir capacities of 0.1–0.6 million m^3 should be built in the lower reaches of the main branch gullies. In the present study, the hydro-sedimentologic process and the geomorphological variation for 18 runoffs were investigated with a middle-scale check-dam of 21 m high, which was built near the outlet of the lower reach of the Yangdaogbou Catchment.

Obviously, deposition on the dam-land is determined by its own reservoir capability as well as the quantity of water-solid mixture from the upper reaches, and is independent of the source of the mixture, as these check-dams are small-sized reservoirs in nature. Thus, the effects of land use, vegetation, rainfall, etc., on soil and water erosions could be embodied in the flux and concentration of the mixture, and the latter could be calculated according to the prototype erosion rate. The Similarity

Theory for deposition in reservoirs with hyperconcentrated flow is practicable and has been applied to several practical projects (Zhang et al. 2001, 2002), so that some of their technical approaches could be used as references in this study.

The experiments were carried out in the Laboratory Hall of the Yellow River Research Center, Tsinghua University, Beijing, China. Due to the limitation of the experimental ground, the length scale λ_L was restricted to 60, which implied that the gully dimension and the dam height were shrunk to 1/60 of the prototype ones. The matrix loess stacked near the experimental spot, which was homologous to the loess in the Yangdaogou Catchment, was used as the erodible material in this experiment. The material had passed through a sieve with 1 cm sieve pores for removing large clods, debris and grassroots. To ensure a regular original microrelief, the soil was prepared by hand-patting for generating a smooth roughness. The diameter of 50% soil particles, d_{50}, was 30.8 μm, and the dry bulk density ρ_d was 1.44×10^3 kg/m³.

No matter it is a field survey or a laboratory experiment, the gravitational effect is more decisive than the viscous one for the turbulent flow caused by rainfall, so the duration of each rainfall is demanded by Froude's Law of Time Scale (Jiang et al. 1994):

$$t_m = \frac{t_p}{\lambda_t} = \frac{t_p}{\sqrt{\lambda_L}} = \frac{10-40}{\sqrt{60}} = 1.3 - 5.2 \text{ (min)} \qquad (8.4)$$

where t is time, min; λ_t is the time scale; λ_L is the length scale. Since the objective of this study is to validate the trend of check-dams in attaining a relative stability, the reliability of the experimental results would not be degraded if several runoffs were performed continuously. Consequently, the duration of each experimental runoff, t'_m, was prolonged to 12 min without making any changes in the sediment concentration and the flow rate, so that it was possible to observe the evolution process of the gully landform in greater details and to complete the experiments more quickly.

In practice, the perennial annual runoff rate and erosion rate are usually taken as design indices for soil and water products in programming the sequence of dam constructing for the check-dam system in small watersheds (Zheng 2003). The flow rate and concentration of the mixture flowing into the reservoir, which were accurately calibrated before the formal experiments, was 10^{-3} m³/s and 80–90 kg/m³, respectively. Thus the soil product pouring into the reservoir in each runoff of the experiment is shown as follows:

$$S'_m = Q'_m \times C'_m \times t'_m = 10^{-3} \times 85 \times 12 \times 60 = 61.2 \text{ (kg)} \qquad (8.5)$$

where S'_m is the soil product, kg; Q'_m is the flow rate of the runoff, m³/s; C'_m is the concentration of the runoff, kg/m³; t'_m is the duration of the runoff, m³/s. All the parameters are measured/calculated in the experiment.

Thus the scale of cumulative sediment for each rainfall can be obtained as follows:

$$\lambda_S = \frac{S_p}{S_m} = \frac{4.29 \times 10^6}{61.2} = 7.01 \times 10^4 \qquad (8.6)$$

Table 8.1 shows the experimental parameters for the runoff simulation experiments. The experimental system was made up of a water circuit and a gully model, as illustrated in Fig. 8.2. Water and loess were added into the round puddling pool and stirred by the spin puddler to become homogeneous enough, and then the mixture was pumped into the constant head tank. To prevent the sediment from subsiding, a reciprocating puddler was set up in the constant head tank. As the sediment-laden

Table 8.1 Experimental parameters for the runoff simulation experiments

Prototype element of the small watershed		Scale	Narration
Name	Amount		
Length of the gully, L (m)	752	$\lambda_L = 60$	Restricted by the experimental ground
Catchment area, A (km^2)	0.206	$\lambda_A = 3265$	Geometrical similarity
Height of the check-dam, H (m)	21	$\lambda_H = 60$	Geometrical similarity
Duration, t (min)	10–40	–	Deforming erosion rate
Silt concentration, C (kg/m^3)	$\approx$200	$\lambda_C = 25$	$\lambda_C = 1.15 - 3$ (Zhang et al. 1994)
Sediment yield (kg)	4.29×10^6	$\lambda_S = 7.01 \times 10^4$	Deforming erosion rate
Dry bulk density, ρ_d (10^3 kg/m^3)	$\approx$1.44	$\lambda_{\rho_d} \approx 1$	Model landform was hand packed

Fig. 8.2 Landscape simulator in which the runoff simulation experiments were conducted

fluid flowed into the gully model, part of it was restored in the upstream of the check-dam and the rest arrived at the storing pool. To prevent the runoff from changing the sediment concentration in the puddling pool, a valve was fixed at the end of the storing pool. Thus, when an experimental run was completed, the valve could be opened, and then the fluid in the storing pool could be released into the puddling pool, which could be reused in the next experiment.

The gully model covered an area of 14 m by 3.5 m. The branch gully at the foreside of the model was "V" shaped, while the main gully at the rear-side was "U" shaped. The length of the branch gully was about 9 m. The dam-site at the main gully was 0.9 m away from the end of the branch, which stood in the entrance of the prototype main gully. The effective height of the dam was 0.35 m before the first runoff. A spillway was located on the left side of the check-dam, which allowed the floodwater to overflow into the storing pool. Among the inlet pipes, an electric flowmeter was fixed to measure the flux invading the gully. The two sidewalls, which were straddled by a measurement bridge with a height finder, were parallel and their tops were of approximately the same altitude. The bridge could move along the sidewalls and the height finder could slide along the bridge, thus the altitude of any point on the gully bed could be measured. The surface altitude was digitized for both horizontal directions in 0.1 m grids.

Eighteen experimental runs were conducted with the same flux of 1.0×10^{-3} m^3/s and the same duration of 12 min. The sediment concentration of the flow running to the gully was kept at approximately 80–100 kg/m^3. There was a 24 h break following each runoff. The initial microtopography made according to the blueprint of the scale model was designated as "$N_m(0)$", the one after the first runoff as "$N_m(1)$", and the one after the second runoff as "$N_m(2)$", and so on. Each microrelief formed in 24 h after the runoff became the initial landform of the next run. Before the next runoff, muddy water stored in the reservoir in the upper reach of the check-dam was defecated and pumped out, and the microrelief of the gully bed was measured by the height finder.

In this experiment, the dam area after each runoff could be calculated by the contour line of the gully bed. After the map of the gully bed had been converted into an AutoCAD file with a ratio of 1 unit to 1 cm, the dam-land enclosed by the contour line could be calculated by using the computer command "Area".

8.3.2 Rainfall Simulation

Rainfall simulations were conducted in the model of a catchment to study the hydro-sedimentologic process and geomorphological evolution in the gullies after the check-dam system was built. The experiments were carried out in the laboratory of the Yellow River Research Center, Tsinghua University, Beijing, China. The landscape simulator consisted of a rainfall simulator suspended above a flume in the small watershed model. Nozzles with an inner rotor were screwed onto the ends of the pipes in the spot, which could spray downwards to the landform 5.5 m below.

The sprinkling intensity could be adjusted by pressure valves to maintain a constant hydraulic head for all nozzles which were located at different elevations. Each basic unit watered a circular area of approximate 5 m in diameter. Two lines of five units each were used to simulate rainfalls in the experimental spot of 6 m by 10.8 m. Clean water was pumped from the nearby reservoir to a constant head tank, and then pumped to the rainfall simulators. A flow-meter and a cut-off valve for each rainfall unit were used to adjust manually the delivery rate of the immersed pump to achieve the desired rainfall intensity. Distribution uniformity of the rainfall was determined using rain-gauges equidistantly spaced over the landscape surface and was measured for three separate periods of ten minutes prior to the experiments. The spatial distribution of the rainfall intensities for the simulator was not uniform, but was constant with time, and was comparable to the sprinkler delivery systems used by others (e.g. Zhou et al. 2000), especially when the rainfall intensity was in the range of 1.0–2.5 mm/min. Drops with an average diameter of about 1.2–2.0 mm, which were measured by catching the drops on a sheet of absorbent paper, were produced by the rainfall simulator (Xu et al. 2004b).

Ten rainfalls were applied after the check-dams were constructed according to the sequence list in Fig. 8.3. Simulated rainfalls resembling a characteristic natural rainstorm in the prototype watershed were applied in the model experiments (Fig. 8.4). For each rainfall, the duration was 20 min, and the intensity was approximate 1.60 mm/min (with errors ≤ 10%). Before the commencement of any rainfall application, the rainfall intensity was verified. The surface of the plot was covered with tarpaulins and the rainfall was collected in a calibrated tank, then measured volumetrically, and was sampled every 2 min. Soil water storage was found to be very sensitive to rainfall (Peugeot et al. 1997), and we had made great efforts to keep

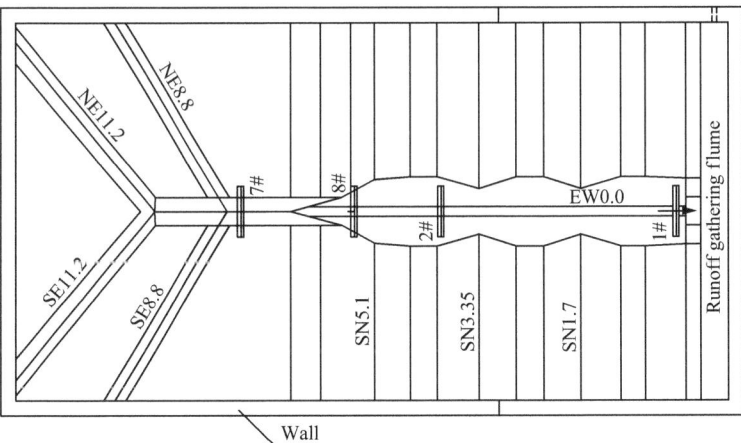

Fig. 8.3 Plane graph of the model relief and arrangement scheme of the check-dams. The check dam arrangement order of schemes A was as follows: dam 7 was constructed before the 2th rainfall session; dam 8 was constructed before the 4th rainfall session; the dam 2 was constructed before the 5th rainfall session; and dam 1 was constructed before the 6th rainfall session

Fig. 8.4 Simulated rainfall was applied on the model check-dam system

an equal antecedent soil moisture before the commencement of the experiment in the same series of rainfall applications. An equal period (24 h) was kept after each rainfall, and the water content was kept almost constant, except in the first experiment. Soil samples and test beds were initially protected by plastic sheets until the rainfall intensity was correctly adjusted and the data logging equipment was operating satisfactorily. Then the cover was removed suddenly and the test was started. During each rainfall application, the total water-solid mixture was collected in the calibrated tank and the runoff container at the outlet of the plot. Runoff samples were also collected with a sampling bottle of 100 ml for determining the sediment concentration. The flow rate was also measured using the calibrated bucket every 2 min. After the experiment, the deposited bed load was dried and weighted, and the weight of the suspended load was calculated from the concentration determined gravimetrically and from the volume measured by the bucket gauge. The microrelief of the gully bed, which was digitized in the horizontal directions in 0.1 m grids, was measured by the height finder. The dam-land area after each rainfall was measured by means of a digital camera and a wooden frame with a known dimension. The wooden frame was placed flatwise on the dam-land, which was proportionally deformed with the dam-land in the plan-photo. Then the areas of the dam-land and the wooden box in the photo could be calculated respectively via the AutoCAD program, and the result so obtained was the practical area of the dam-land.

Before the above scenario demonstration as the dams were built orderly, the ratio of the erosion extents between the prototype and the model for each run of rainfall should be verified. An identifying test of simulated rainfall before the construction of the dams, in which the geometrical scale, the antecedent land cover and the rainfall in the identifying test were homologous to those of the above experiment for the dams,

Table 8.2 Experimental parameters for the rainfall simulation experiments

Prototype element of the small watershed		Scale	Narration
Name	Amount		
Length of the gully, L (m)	3008	$\lambda_L = 240$	Restricted by the experimental ground
Height of the check-dam, H (m)	*	$\lambda_H = 240$	Geometrical similarity
Catchment area, A (km^2)	3.3	$\lambda_A = 57,600$	Geometrical similarity
Duration, t (min)	120–480	$\lambda_t = 7.75$	Gravity similarity
Silt concentration, C (kg/m^3)	≈200	$\lambda_C \approx 3$	$\lambda_C = 1.15 - 3$ (Zhang et al. 1994)
Sediment yield (kg)	6.87×10^7	$\lambda_S = 5.65 \times 10^5$	Deforming erosion rate
Dry bulk density, ρ_d (10^3 kg/m^3)	≈1.56	$\lambda_{\rho_d} \approx 1$	Model landform was hand packed

*Dam 7 is 35 m high, and Dams 1, 2 and 8 are 68 m high

was conducted. The cumulative sediment volume for the experiment is 121.53 kg. Thus the scale of cumulative sediment for each rainfall can be obtained as follows:

$$\lambda_S = \frac{S_p}{S_m} = \frac{6.87 \times 10^7}{121.53} = 5.65 \times 10^5 \tag{8.7}$$

where S is the soil production, kg; the subscript p represents the prototype; the subscript m represents the model. The parameters of the rainfall experiments are presented in Table 8.2.

8.4 Results and Discussion

The decrease in dam-land elevations and the re-stabilizing of the gully gradients after the building of the check-dams represent a hydro-sedimentologic balance of the check-dam or the check-dam system, and the building of the check-dam is an effective way to control soil erosion in the gullies.

Mean dam-land altitude variation. Both in the runoff simulation experiment for a single check-dam and the rainfall simulation experiment for the check-dam system, the increase in dam-land altitude became smaller and the gully gradient became more stable as the experiment was progressing. Figure 8.5a illustrates that the increase in dam-land altitude for each runoff became gradually smaller in the runoff experiment. As the water-solid mixture was flowing to the check-dam during the experiment, the thalweg was eroded in the upper reach of the gully, which was out of control of the check-dam, and the gully-bed in the lower reach was filled up, and even a certain land area was gradually formed in the upper reach of the dam. Although

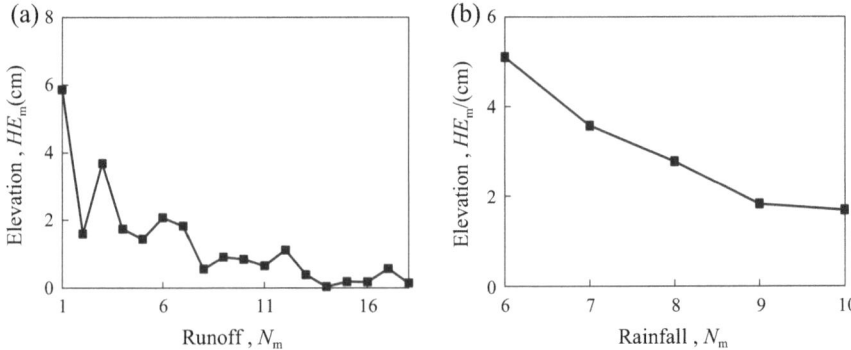

Fig. 8.5 Variation of the mean dam-land altitude with runoff/rainfall. **a** Single check-dam; **b** Check-dam system

the altitudes of the dam-land after each runoff had shown an increasing trend, the increased amount before the $N_m(12)$, especially before the $N_m(5)$, was greater than that after the $N_m(13)$. In the rainfall simulation experiment, reservoirs of Dam 2, 7 and 8 were filled up when one or two runs of simulated rainfall had been conducted, due to the too exquisite erosion extent of each rainfall. However, Dam 1 was not filled up and the dam-land was growing up all along owing to its larger capacity and the time of construction was late. The elevation trend of this dam-land was analogous to that of the single check-dam in the runoff simulation experiment, as shown in Fig. 8.5b.

Mean slope gradient variation. The base level of the gully was raised and the mean gradient was lowered as depositions took place on the dam-land. Referring to ASL (Casalí et al. 1999), we defined PL as the length-weighted average slope gradient:

$$PL = \frac{\sum L_i P_i}{\sum L_i} \tag{8.8}$$

where L_i is the length of each segment of the gully, and P_i is the slope gradient of this segment with uniform characteristics. In these experiments, L_i and P_i could be calculated based on the coordinates along the gully thalweg. PL of the main gully after each runoff/rainfall simulation experiment for a single check-dam and a check-dam system was listed in Fig. 8.6, respectively. The gradient was steeper in the initial stages, but became less steep as the experiment went on, since the dam-land area was enlarged. For the check-dam system experiment (Fig. 8.6b), the gradient of the main gully was becoming gentler, as Dam 1 had not been filled up yet even after 10 simulated rainfalls. However, in the runoff experiment of the single check-dam (Fig. 8.6a), beginning from the $N_m(13)$ runoff, the PL was kept at 7%, and almost did not vary as the experiment during the experiment. This means that the gully had become relatively stable, and the slope had only varied a little, despite the minute variations in the microtopography.

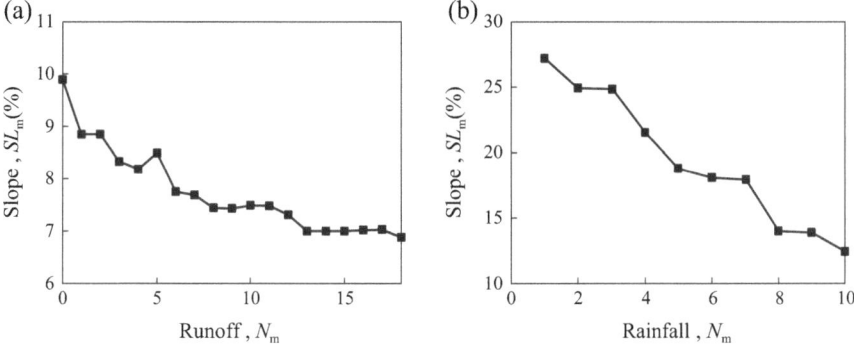

Fig. 8.6 Variation of the mean slope with runoff/rainfall. **a** Single check-dam; **b** Check-dam system

However, if the runoff experiments were continued, the increasing rate in dam-land elevation would become much slower. It was really difficult to provide strict threshold values for gully slopes and dam-land elevations only by model experiments. In practice, the threshold value of dam-land increase per year which could be borne by the local peasants is decided by the local social and economic conditions (Li 2004), and generally amounts to 0.3 m/a (Zheng 2004; Wang and Ma 2003).

In fact, from our experimental results, the increase rate of the dam-land altitude was evidently shrinking as the runoff was carrying on, and the trend was exhibiting a relative stability in the gully in respect of the hydro-sedimentologic balance.

8.4.1 Mechanism of the Relative Stability Development

Figure 8.7 shows the variation in composition of the sediment on the surface of the gully bed after 18 runs of model runoff. Just like any other water-blocking structure

Fig. 8.7 Coarsening of the gully bed after runoffs

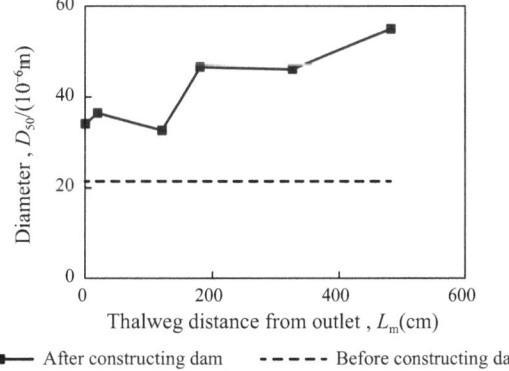

such as reservoirs, retention ponds, etc., the check-dam can slow down the velocity of the sediment-laden water, resulting in a lower sediment transport capacity. Thus, some of the sediments would sink, and the sediment concentration in the outlet would become smaller than that in the inlet. Moreover, before the next runoff, the sediment-laden fluid stored in the reservoir would become clear, and some sediment would be left on the dam-land to prevent the dam-land from erosion.

On the other hand, development of a "shielding" layer composed of relatively heavy soil particles could protect the underlying soil from runoff erosion. When the soil surface was initially inundated with runoffs, the soil particles were detached from the soil surface and were entrained into the gully flow. Lighter particles with low settling rates would move far away from their original locations, whilst heavier particles would settle more quickly near their original positions. If this process was continued, eventually most of the lighter particles would be removed, leaving a shielding layer of heavier particles, which could act to protect the underlying soil.

As the dam-land rose, its area expanded in virtue of the sloping gully bank. Figure 8.8 illustrates a linear rising trend of the dam-land area. In the rainfall simulation experiment for the check-dam system (Fig. 8.8b), Dams 2, 7 and 8 had been filled up before the 6th rainfall simulation. The increase in dam-land area in the figure was mainly attributed to Dam 1. Whether it was in the single check-dam experiment or in the check-dam system experiment, the dam-land areas were expanding linearly, as shown in Fig. 8.8. Consequently, even if the bulk of the intercepted sediment was unvaried for every rainfall, the dam-land would rise to a smaller extent owing to the augmenting of the area after each runoff.

However, other elements can also contribute to the relative stability of the check-dams. For example, as the dam-land rises, the reservoir capacity will decrease, and less sediment-laden fluid can be stored in the upstream of the check-dam. As a result, the amount of deposited sediment will decrease as well. In fact, in the later stage of integrative control of small watersheds of the Chinese Loess Plateau, the dam-land has increased more slowly owing to a great reduction of the amount of sediment

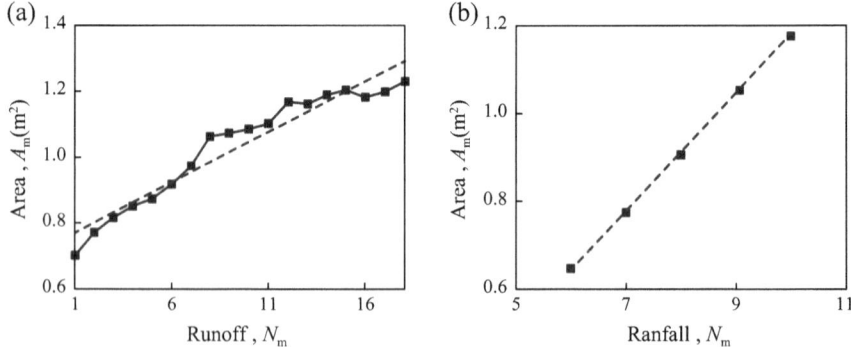

Fig. 8.8 Variation of the alluvium area with runoff/rainfall. **a** Single check-dam; **b** Check-dam system

originating from the slopes and gullies, thus causing the smaller magnitude of floods and smaller sediment concentrations.

Scour/deposition balance is a common phenomenon of gully evolvement which has been testified by investigations on other watersheds. The rate of gully incision is controlled by the flow rate, depth, turbulence and temperature of the water flow, as well as by soil texture, soil mechanical pattern, and the level of vegetation protection (Sidorchuk 1999). Foster (1982) has found that gullies were eroded downwards, until a layer of low erosion susceptibility was reached. After this, the gullies would become widened, and the erosion rate would decrease. Eventually, the gullies would "equilibrate" with the overland flow. Novak (1985) has developed equations for calculating the time required to reach equilibrium for rill and channel erosions. Recent studies have indicated that gully formation can be divided into two stages (Sidorchuk 1999). The initial stage takes up about 5% of a gully's lifetime during which the morphological characteristics of a gully, such as length, depth, width, area and volume, are developing, and stability cannot be attained. After this, a stable stage will be reached, and the gully will attain a maximum in size. In the initial stage, hydraulic erosion is predominant at the gully bottom, and rapid mass movements occur on the gully sides, whilst in the stable stage, sediment transport and sedimentation become the main processes at the gully bottom, and the gully width increases limitedly due to lateral erosion. The mass movements go on slowly in this stage. In our natural world, if the local conditions are permitting, some streams can maintain their relatively stable channels over even a century, despite that there have been great floods in such a long period of time (Warburton et al. 2002; McEwen 1994; Warburton et al. 1993).

As check-dams have been built in small watersheds on the Loess Plateau, the relieves of the gullies were changed. Substantial sediment caused by rainfalls and floods have been stored on the dam-land for decades, and the mean slope gradients of the gullies became gentler. Consequently, erosion in the gullies has been alleviated, and the gullies have attained a relative stability at last. On the other hand, even though the amount of sediments from the upper reaches has not been reduced, the rising rate of the dam-land becomes slower due to the enlargement of the dam-land area. Hence a critical height of a check-dam exists for a given gully having definite conditions of soil, water, geology and physiognomy. If the check-dam in a gully is higher than the critical value, and there are depositions in the upper reach of the check-dam, the mean slope gradient of the gully would be gentle enough to diminish the erosion greatly. As a result, if the annual sedimentation thickness is very little in the later stage of the dam-land formation, the workload for heightening the dam is so little that it can be afforded by the local people (Zeng et al. 1999). Although other factors also play their roles in maintaining the relative stability of the check-dam systems (Fang 1995), a equilibrium of soil and water in the gullies is the most important criterion.

In fact, many natural check-dams, such as the Balihe Gully and the Sanshilihe Gully in Shannxi Province and the Laobatou and the Qianqiuzi in Gansu Province, all have achieved a good relative stability. Huangtuwa in the Zizhou County of Shannxi Province, another famous natural check-dam covering 40 hm^2 of lowlands, has been working favorably for more than 400 years. Some check-dam systems built in the 1960s, such as the Kanghegou check-dam system in Shanxi Province and the Wangmaogou dam system in Shannxi Province, are also typical projects attaining good relative stabilities, and have retained enormous amounts of sediments and yielded plentiful crops. Hence, it is most evident that the attaining of relative stability should be the ultimate goal to build check-dam systems.

8.4.2 Further Research on the Theory of Relative Stability

The amount of dam-land increase per year. It is supposed that floodwater was evenly spread on the dam-land in the initial hypothesis on the relative stability of check-dams. In fact, the water depth near the dam is larger than that far from the dam, due to the longitudinal deposition profile existing in the upper reach of the check-dam, and the wallop of the floodwater pouring into the reservoir. According to the field investigation, the average longitudinal slope gradient of the check-dam lands in Shannxi Province is 0.2–0.5%, which makes an obvious impact on the distribution of the impounded floodwater, especially that on the check-dam land with large reservoir capacity (Tian et al. 2004). The phenomenon mentioned above was also widely found in the rainfall simulation experiments. Figure 8.9a shows the slope gradient of the check-dam land gradually became small in the runoff experiments of the single check-dam. Figure 8.9b shows the change of the gully bed after 9 events of rainfalls. Both figures illustrate that the longitudinal slope gradient of the check-dam land was larger before the reservoir filled, and it became smaller after the reservoir filled, but the gradient never be equal to 0.

Consequently, the annual increment of the check-dam height in the connotation of relative stability should be the spatial mean value of the dam-land increase per year (Fang 1995). Experimental data related to the amount in this chapter had been also processed according to the above method. Clarification of the definition is helpful to construction of the check-dam systems in the practice of erosion control.

The relationship between the construction of the check-dam system and the control of the slope erosion. From the previous experimental data it could be found that the main function of the check-dam system is to control soil and water loss in the gullies, other than directly alleviate soil and water loss on the slope. Thus comprehensive treatment including tree planting, terrace constructing and pit digging should be implemented order to improve the eco-environment of the Loess Plateau.

Actually, the relative stability of the check-dam system and the control of the slope erosion may be mutually promoted. Gully is usually the influx of the runoffs, which is also the channel of surface runoffs, therefore, the water resources are comparatively centralized. With the reservoir of the check dam system in the gully, the water resources could be fully impounded to irrigate the plant on the check-dam and the slope, and indirectly reduce soil and water loss on the slope. At the same time, the control of slope erosion could reduce the flood peak to a certain extent, which favors the flood control of the check-dam system. Besides, the control of the slope erosion decreases the amount of soil flowing to the reservoir. Consequently the

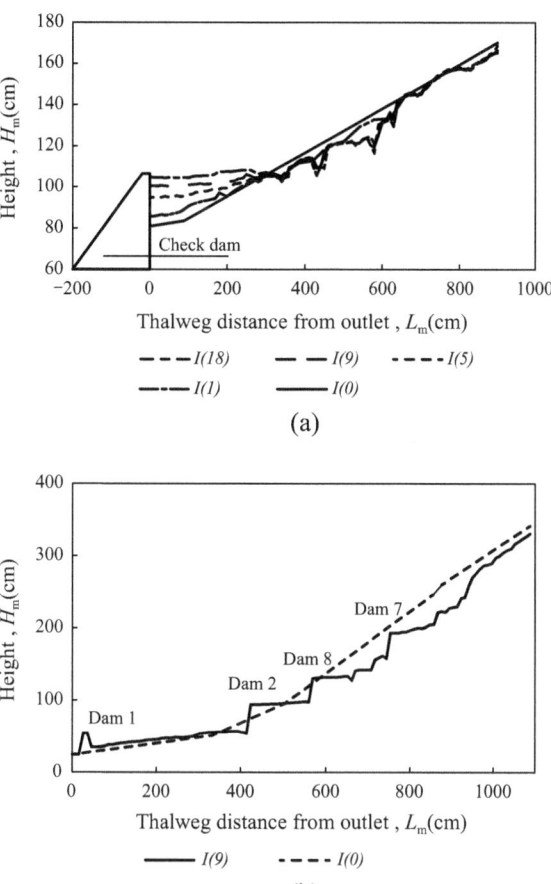

Fig. 8.9 Longitudinal deposition profile of the check-dam land. **a** Single check-dam; **b** Check-dam system

annual increase of the check-dam land becomes smaller, and the equilibrium of soil and water in the catchment would be easy to be realized.

Relationship between relative stability of the dam system and that of the single dam. The relative stability of the dam system was firstly recognized from the natural Juqiu on the Loess Plateau, and then the relative stability of the check-dam system was developed to emphasize the safety of the flood preventing and controlling (Fang 1995). Unfortunately, a big dispute arose on the relationship between the critical coefficient for relative stability of the dam system and the flood preventing and controlling. As stated before, the relative stability of the dam system mainly refers the balance of the soil and water. Relative stability of the check-dam system has positive effects on the safety of the dam system, nevertheless, it is not the determined factor in the structural safety of a single check-dam.

The relative stability of the dam system reflects the overall development of the check-dam system. The premise is that the soil and water production on the whole slope is homogeneous, and the deposition and inundation on a single dam is also basically identical (Zhu et al. 1997). Therefore, the relative stability of the dam system only represents the relative balance of the soil and water in some limited small and ideal catchment (Shi 2005). Up to the present, the dam system and single check-dam are not clearly separated in the literatures on the balance of the soil and water. The concept of relative stability mainly replies to the relative balance of soil and water retained by a single check-dam.

In combination with the model experiment of the dam optimization programming, we performed the rainfall simulation experiments with several scenarios of dam construction sequence and found out that the small dams in the reservoir were quickly filled up.

The most contribution of the relative stability of the dam system is that the concept of the critical coefficient for relative stability provides a quantitative criterion for the dam system programming in the small watershed, with which the number of dams could be figured out according to the corresponding reservoir capability.

8.4.3 Comparison of Sediment Loss and Dam-Land Area

The main purposes of the check-dam system design are to (1) retain sediment and (2) form dam-lands. The first objective, which is the most important for the total area of dam-lands, is determined by the retained sediment (BMUM 2004). In fact, the effects of sediment-storage are determined by the order in which check-dams are constructed. It is preferable to build the larger reservoirs with greater flood-control capabilities in the lower reaches of the main gully (e.g., Scheme A here), rather than building dams in the upper reaches first (e.g., Scheme B here).

The scheme with less cumulative soil loss is better, as the antecedent land cover and the amount of rainfall is the same in both schemes. The amount of soil loss after each rainfall event is shown in Table 8.3. The total amounts of sediment retained by the dams in both schemes were obviously different. The total amount of soil loss in Scheme B was 13.5×10^6 kg after 10 rainfall events, and only 10.6×10^6 kg in Scheme A, 27% smaller than that in Scheme B. Especially in the first rainfall event, sediment was impounded more effectively and much less soil was lost in Scheme A than in Scheme B (Fig. 8.10). Thus, the order of dam construction in Scheme A is preferable, for less sediment may escape to the Yellow River than in Scheme B. Moreover, the dam system attains relative stability earlier than in Scheme B, thereby increasing the benefit of the check dam System (Xu et al. 2004a).

The dam-land area is generally determined by the gully shape and local social and economic conditions. After the dam has been built, the area of the deposited dam land will rise and the gully erosion will decrease as time goes on. As a result, a better benefit of the check dam system may be received. Since the third rainfall event, the cumulative area of the dam-land in Scheme A was larger than that in Scheme B, as shown in Fig. 8.11. After 10 rainfall events, the cumulative area of the dam-land was 23 hm^2 as the check-dams were built according to Scheme A, whereas the cumulative

Table 8.3 Soil and water loss in the rainfall simulation experiments

Rainfall event	Scheme A			Scheme B		
	Dams built before the rainfall event	Sediment loss, S_p (10^6 kg)		Dams built before the rainfall event	Sediment loss, S_p (10^6 kg)	
		Individual	Total		Individual	Total
1	Dam 1	0.7	10.6	Dam 9, 10, 11 & 12	4.1	13.5
2		2.5		Dam 7	3.9	
3	Dam 2	1.3		Dam 3, 4, 5 & 6	1.9	
4	Dam 3, 4, 5 & 6	0.8		Dam 8	1.4	
5	Dam 7	0.9		Dam 2	0.9	
6	Dam 8, 9, 10, 11 & 12	0.9		Dam 1	0.3	
7		0.7			0.3	
8		0.9			0.3	
9		1.1			0.3	
10		0.7			0.2	

Notes Dams 1, 2 and 8 are 18 m high, Dams 3, 4, 5, 6 and 7 are 9 m high, and Dams 9, 10, 11 and 12 are 5.4 m high, respectively

Fig. 8.10 Schemes A versus B: a comparison of the cumulative amounts of the soil losses

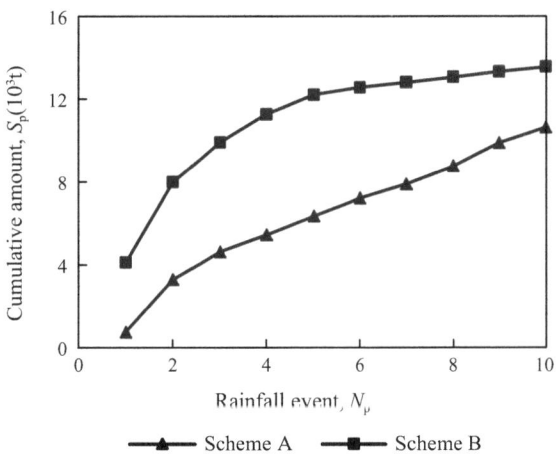

area was only 19 hm^2 according to Scheme B. Hence Scheme A was preferable to Scheme B based on the dam-land area. Although the cumulative amount of soil losses and dam-land areas would eventually be quite similar for Scheme A and Scheme B after enough rainfalls, the dam-land benefit to farming would be more rapid, and the measures to control soil erosion would take effect more quickly when check dams were constructed following Scheme A instead of Scheme B.

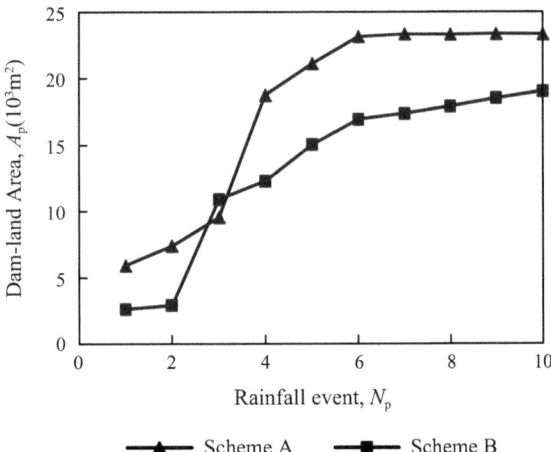

Fig. 8.11 Schemes A versus B: a comparison of the cumulative areas of the dam-lands

8.4.4 Comparison of the Water Discharge and Sediment Concentration

The effects of the construction schemes on runoff and sediment concentration are shown in Figs. 8.11 and 8.12. The average discharge for each rainfall in Scheme A was close to the corresponding discharge for the same rainfall event in Scheme B (as shown in Fig. 8.12), so did the total amount of water loss after 10 rainfalls (Table 8.3).

However, the order of dam construction had a significant effect on sediment concentration of the mixture through the outlet, for the soil concentration of the mixture from the upper reaches into the outlet caused by the early runs of rainfall was higher than that by the latter runs, and consequently, more soil was left behind on the dam

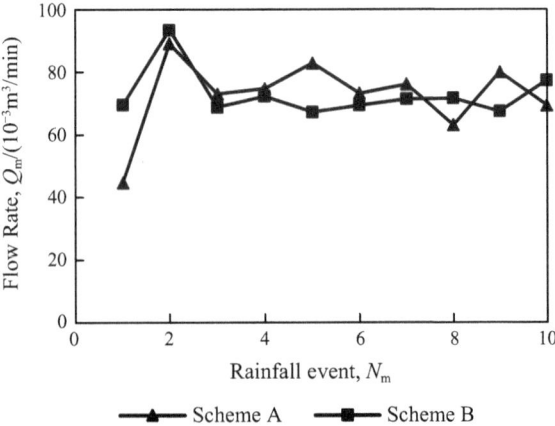

Fig. 8.12 Comparison of flow rate between Scheme A and Scheme B

land. Thus, the amount of retained soil was positively correlated with the earliness of construction of the key dams.

The soil concentrations of the runoff from the outlet in Scheme A were less than those for the corresponding rainfall events in Scheme B for the first 4 rainfall events. Especially during the first rainfall event, the average concentration was $25.32\,\text{kg/m}^3$, which was only 14.5% of that in Scheme B, $174.39\,\text{kg/m}^3$ (as shown in Fig. 8.13). For Scheme A, Dam 1, which was built near the outlet of the watershed with the largest reservoir capacity, had retained a large amount of sediment. Thus, the impounding effect was remarkable especially during the initial stage of check-dam building. For Scheme A, the total amount of soil loss was 40.89 kg in the first simulated rainfall event, which was only 6.9% of the total amount of the retained sediment. However, as to Scheme B, the volume of sediment loss was 223.10 kg in the first simulated rainfall event, accounting for 30.3% of the total amount of sediment retained. The amount of runoff during the first rainfall event retained in the reservoir in Scheme A was larger than that in Scheme B, because Dam 1, which had the largest reservoir capacity, was built before the first rainfall event in Scheme A. Comparatively, in Scheme B, the advantage of large reservoir capacity for Dam 1 was not fully taken as Dams 2 to 12 were all built before Dam 1. Dam 1 with the largest capacity was built before the sixth rainfall when the sediment concentration in the runoff entering the reservoir was decreasing. As a result, more soil was lost after 10 rainfalls in Scheme B than in Scheme A.

Fig. 8.13 Comparison of the mixture concentration in the outlet between Scheme A and Scheme B

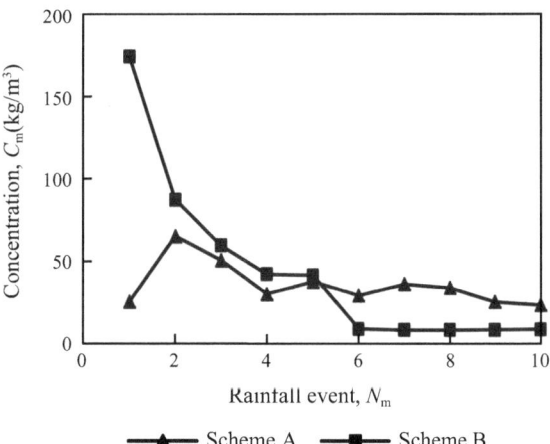

8.4.5 Relationship Between the Dam-Land Increase and Erosion Reduction

Typically, the amount of gully erosion exceeds that of the interrill erosion. The gully erosion would be alleviated if the gullies with serious gravity erosions become a flat dam-land due to sediment deposition. As we know, soil erosions varied in different regions in the small watershed. Erosion control could be improved when the area with serious soil loss is controlled first. Compared with the order in Scheme A, the order of building the check dams in Scheme B was "from the upper reaches to the lower reaches" and "from the branch rill to the main gully". Thus, soil loss in the lower reaches of the main gully and branch rills with the most serious erosion was not completely controlled, and enormous soil was lost. In this experiment, the sum of sediment loss in the first two rainfall events was 60% of the total after 10 rainfall events in Scheme B, while only 31% in Scheme A.

On the other hand, after the check dam system has been constructed, the soil was retained in the upper reaches of the check dams, the dam-lands were enlarged, and the gravity erosions on the gullies were reduced. Nevertheless, the forming rates of the dam-land were different as different sequences of building dams were adopted even though the antecedent land covers and check dams were the same. When the check dams were built following Scheme A, all check-dams in the main gully were filled after six rainfalls. In contrast, when the check dams were built following Scheme B, Dam 1 was not filled after 10 rainfall events. Therefore, farming work would be delayed and the economic benefit would be reduced due to the delayed formation of the dam-land.

8.5 Conclusions

Hydro-sedimentologic balance is the most important element for realizing relative stability and optimal planning of the check-dams, which means that soil and water in the small watersheds controlled by the check-dams could be internally absorbed, without the need of raising the height of the dam. Quantitative results from the experiments indicate that the gullies could auto-stabilize themselves after the check-dams were built, as raising of the dam-land altitude became slower and the mean gradient of the gully was kept at a constant. The experiments also show that the sequence of constructing dams from the lower reaches to the upper reaches was preferable, for less soil will be escaped from the catchment.

References

Albertson M L, Barton J R, Simons D B. 1960. Fluid mechanics for engineers. Prentice-Hall, Englewood Cliff, NJ: 561.

BMUM (Bureau for Management of the Upper and Middle Reaches of the Yellow River). 2004. Programming for Check-dams. Beijing: Chinese Project Press: 80–81 (in Chinese).

Cai Q G, Wang G P, Cheng Y Z. 1998. The process and simulation of soil erosion for small watersheds on the Loess Plateau, China. Beijing: Science Press: 69–72, 135–137, 177, 189–190 (in Chinese).

Casalí J, López J J, Giráldez J V. 1999. Ephemeral gully erosion in southern Navarra (Spain). Catena, 36(1–2): 65–84.

Fang X M. 1995. Criterion and condition for the relative stability of check-dam system. Soil and Water Conservation in China, (11): 29–32 (in Chinese).

Fang R Y. 1996. Consideration on check-dam system agriculture in Shanxi Province. Soil and Water Conservation in China, (7): 47–49 (in Chinese).

Fang X M, Wang Z H, Kuang S F. 1998. Mechanism and effect of check-dam to intersect sediment in the middle reach of Yellow River. Journal of Hydraulic Engineering, (10): 50–54 (in Chinese).

Foster G R. 1982. Modeling the erosion process// Hann C T, Johnson H P, Brakensiek D L. Hydrologic modeling of small watersheds. Monograph No.5 American Society of Agricultural Engineers, St. Joseph. Michigan: 297–380.

Hancock G R, Willgoose G R. 2004. An experimental and computer simulation study of erosion on a mine tailings dam wall. Earth Surface Processes and Landforms, 29(4): 457–475.

Jiang D S, Zhou Q, Fan X K, et al. 1994. Simulated experiment on normal integral model of water regulating and sediment controlling for small watershed. Journal of soil and water conservation, 8(2): 25–30 (in Chinese).

Lei Y J, Zhu X Y. 2002. Formation, control principle and method for relative stability of check-dam system. Yellow River, 22(2): 23–26 (in Chinese).

Li M. 2004. Analyse on relative stability for check-dam system// Science and Technolocy Spread Center of water resource ministry and Yellow River Research Society. Proceedings of Key Technique to Construct Check-dam System in the Small Watershed on the Loess Plateau, China. Xi'an: 5–9 (in Chinese).

Li J, Qin X Y, Liu L W. 1995. A brief discussion on method of planning for comprehensive small watershed control in China. Bulletin of Soil and Water Conservation, 15(3): 8–11,32 (in Chinese).

Lin M H, Zhu M X, Bai F L, et al. 1995. Optimization planning model of small watershed dam system and its application. Yellow River, (11): 29–33 (in Chinese).

McEwen L J. 1994. Channel planform adjustment and stream power variations on the middle River Coe, Western Grampian Highlands, Scotland. Catena, 21(4): 357–374.

MWRC (Ministry of Water Resources, P.R.C). 2003. Programming for Check-dams in Loess Plateau (Technical Report): 47–48 (in Chinese).

MWRC (Ministry of Water Resources, P.R.C). 2009. Regulation of techniques for comprehensive control of soil erosion- Technique for erosion control of gullies. GB/T 16453.3—2008. Beijing: China Water Power Press (in Chinese).

Novak M D. 1985. Soil loss and time to equilibrium for rill and channel erosion. Transactions of the ASAE, 28(6): 1790–1793.

Peugeot C, Esteves M, Galle S, et al. 1997. Runoff generation processes: results and analysis of field data collected at the East Central Supersite of the HAPEX-Sahel experiment. Journal of Hydrology, 188–189(96): 179–202.

Shi X J. 2005. Development and suggestions on relative stability of the check-dam system in the small watershed of the Loess Plateau. China Water Resources, (4): 49–50 (in Chinese).

Shi H, Tian J L, Liu P L, et al. 1997a. Study on spatial distribution of erosion yield in a small watershed by simulation experiment. Research of Soil and Water Conservation, 4(2): 75–84,95 (in Chinese).

Shi H, Tian J L, Liu P L. 1997b. Study on relationship of slope-gully erosion in a small watershed by simulation experiment. Journal of Soil Erosion and Soil and Water Conservation, 3(1): 30–33 (in Chinese).

Sidorchuk A. 1999. Dynamic and static models of gully erosion. Catena, 37(3–4): 401–414.

Tian Y H, Fu M S, Mu Z L, et al. 2003. Key technology to construct the check dam system. China Water Resources, (17): 59–61 (in Chinese).

Tian Y H, Liu H J, Yang M, et al. 2004. Relative stability condition and feasibility of the check-dam system in the small watershed of the Loess Plateau// Scientific Extension Center of Water Conservancy Department and Association for Yellow River Research. Proceeding of the Workshop on Key Technology Check-Dam System Construction in the Small Watershed of the Loess Plateau. Zhengzhou: 21–24 (in Chinese).

Timmons M B. 1984. Use of physical models to predict the fluid motion in slot-ventilated livestock structures. Transactions of the ASAE, 27(2): 502–507.

Wan S C, Lin L X, Wan Y Q. 1995. A nonlinear programming model for study on optimization planning of a dam system. Journal of Wuhan University of Hydraulic and Electric Engineering, 28(3): 260–266 (in Chinese).

Wang Y S, Ma H. 2003. Research and application on the relative stability coefficient of check-dam system. China Water Resources, (17): 57–58 (in Chinese).

Warburton J, Davies T R H, Mandl M G. 1993. A meso-scale field investigation of channel change and floodplain characteristics in an upland braided gravel-bed river, New Zealand. Geological Society, London Special, Publications, 75(1): 241–255.

Warburton J, Danks M, Wishart D. 2002. Stability of an upland gravel-bed stream, Swinhope Burn, Northern England. Catena, 49(4): 309–329.

Wu Y C. 1994. Synchronous optimization among silt arrester system, dam height and damming period with region-varied linear programme. Journal of Soil Erosion and Soil and Water Conservation, 8(4): 60–65,90 (in Chinese).

Wu Y C, Huang L. 1995. Existing condition of the best time of key dam system construction and calculation of actual silting-up date. Soil and Water Conservation in China, (6): 21–24,62 (in Chinese).

Xu M, Wang G. 2000. To accelerate the construction of check-dams in the Loess Plateau. Yellow River, 22(1): 26–28 (in Chinese).

Xu X Z, Zhang H W. 2004. A method to design scaled model experiment for dam programming in the small watershed of Loess Plateau, China// The Yellow River Conservancy Commission. Proceedings of Model based Loess Plateau. Xi'an: 142–149 (in Chinese).

Xu X Z, Zhang H W, Zhang O Y. 2004a. Development of check-dam systems in gullies on the Loess Plateau, China. Environmental Science and Policy, 7(2): 79–86.

Xu X Z, Zhang H W, Zhu M D. 2004b. Study on measuring method of particle size of raindrop and its improvement. Soil and Water Conservation in China, (2): 22–24 (in Chinese).

Xu X Z, Zhang H W, Xu S G, et al. 2009. Effects of dam construction sequences on soil conservation efficiency of a check-dam system. Journal of Beijing Forestry University, 31(1): 139–144 (in Chinese).

Yuan J P, Jiang D S, Gan S. 2000a. Simulated experiment on normal integral model of different control degrees for small watershed. Journal of Natural Resources, 15(1): 91–96 (in Chinese).

Yuan J P, Lei T W, Jiang D S, et al. 2000b. Simulated experimental study on normalized integrated model for different degrees of erosion control for small watersheds. Transactions of the Chinese Society of Agricultural Engineering, 16(1): 22–25 (in Chinese).

Zeng M L, Fang X M, Kang L L, et al. 1995. Relative steadiness of development of valley-dam system is absolutely feasible. Yellow River, (4): 18–21 (in Chinese).

Zeng M L, Zhu X Y, Kang L L, et al. 1999. Effects of sediment reduction and erosion control and development prospects of warping dam in water and soil loss areas. Research of Soil and Water Conservation, 6(2): 126–133 (in Chinese).

Zhang H W. 1994. The study of the law of similarity for models of flood flows of the lower reach of the Yellow River. Beijing: Tsinghua University (in Chinese).

Zhang H W, Jiang E H, Bai Y M, et al. 1994. Similarity law for physical model of the hyperconcentrated flow in Yellow River. Zhengzhou: Henan Science and Technology Press: 80, 115, 156–162 (in Chinese).

Zhang Z G, Wang G P, Jia Z J, et al. 1995. Sediment source of Wangjiagou Gully in a well-controlled way. Science and Technology on Soil and Water Conservation in Shanxi, (2): 6–8 (in Chinese).

Zhang J H, Zhang H W, Jiang C B, et al. 2001. Primary study on physical model similarity for reservoir on the Yellow River. Journal of Hydroelectric Engineering, (3): 53–58 (in Chinese).

Zhang H W, Zhang J H, Wang G D, et al. 2002. Design of the movable bed model for the Dongzhuang reservoir on the Jinghe River. Journal of Sediment Research, (1): 71–77 (in Chinese).

Zheng X M. 2003. Discussion on dam-system construction on the Loess Plateau of China. China Water Resources, (17): 19–22 (in Chinese).

Zheng B M. 2004. Discussion on key technology to construct check-dam systems in the small watershed. Proceedings of Key Technique to Construct Check-dam System in the Small Watershed on the Loess Plateau, China Science and Technology Spread Center of Water Resource Ministry and Yellow River Research Society. Xi'an: 16–20 (in Chinese).

Zhou P H, Zhang X D, Tang K L. 2000. Rainfall installation of simulated soil erosion experiment hall of the state key laboratory of soil erosion and dryland farming on Loess Plateau. Bulletin of Soil and Water Conservation, 20(4): 27–30,45 (in Chinese).

Zhu X Y, Lei Y J, Liu L B. 1997. Cognition on several important problems about relative stablility of dam system. Soil and Water Conservation in China, (7): 53–55 (in Chinese).

Chapter 9
Gravity Erosions on the Loess Gully Bank: Avalanche, Landslide, or Mudslide

Abstract Gravity erosion is a dominant geomorphic process on the steep loess slopes. Here, we conducted rainfall simulation experiments to monitor occurrence and behavior of the mass failure on steep loess slopes. The results show that the quantity of soil loss caused by the avalanche and landslide was much more than that caused by the mudslide, and the avalanche was the most violent gravity erosion. As the slopes were eroded with five runs of rainfalls each at a duration of 48 mm, the total volume of avalanche, landslide, and mudslide were 150.9×10^3, 82.5×10^3, and 3.9×10^3 cm^3/m, and accounted for 62, 36, and 2% of the total gravity erosion, respectively. Furthermore, the slope height and gradient had a remarkable positive correlation with the erosion amount. As a result, the avalanche and landslide, especially the former, played a crucial role in soil erosion on steep slope compacted by hand with loess.

Keywords Mass failure · Behavior · Role · Trigger · Loess Plateau

9.1 Impact Factors of the Gravity Erosion

Gravity erosion, mass failure on steep slope triggered by self-weight, is an important process controlling the sedimentary structures and growth patterns of the steep slope, and also one of the major sources causing a large amount of soil loss to the lower reaches. On the Loess Plateau of China (Fig. 9.1), soil and water loss as well as gravity erosion is the severest in the world. How do we quantitatively evaluate the roles of various mass failures on the steep slope, and availably alleviate gravity erosion according to local characteristic conditions? The questions above have important implications for modeling soil erosion in the small watershed and it is also significant in controlling failure disasters.

Gravity erosion is a frequent and widespread geomorphological phenomenon, whether in mountainous or urban areas. It is the mass failure on a steep slope, triggered by self-weight. Erosion due to gravitational force occurs under the combined influence of definite hydrologic, geologic, and topographic conditions. Gravity erosion is also an important part of the loss process and is often the first stage in the

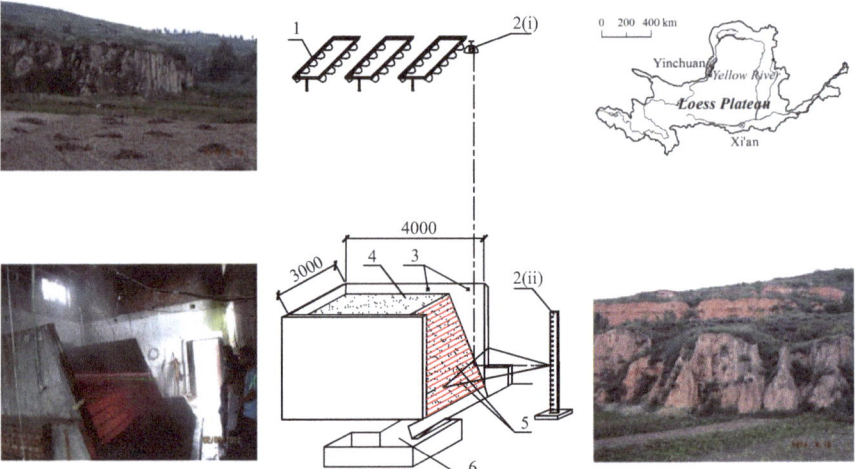

Fig. 9.1 Study area and experimental arrangement. Keys: 1. Rainfall simulator; 2. Topography meter (i: Camera with a collimator, ii: Laser source); 3. Positioning marks; 4. Model slope; 5. Equidistant horizontal projections; 6. Receiving pool

breakdown and transportation of weathered materials. The phenomena can be classified in part by the spatial size and distribution on the ground, or by the duration of time that the process acts, or by the rate of movement (Wang et al. 2014; Shroder and Bishop 1998). It may also differ with respect to the thickness of a failed mass, time of failure occurrence, or rotational inclination (Au 1998). Forms of gravitational erosion include avalanche, landslide, mudflow, and sinkhole formation. Climate and landform play significant roles in the occurrence and behavior of gravity erosion. Gravity erosion generally takes place together with hydraulic erosion, namely, soil loss due to water flowing over the slope, but the mechanism and dynamics of each type of erosion are different. Hence the measures to control hydraulic erosion and gravity erosion are different, and it is essential to quantitatively distinguish the amounts of the failure masses during the same event of rainfall (Xu et al. 2015).

Most gravity erosion occurs during or just after storms (Ali et al. 2014; Peruccacci et al. 2012; Tsai and Yang 2006; Salciarini et al. 2006; Fourie 1996; Montgomery and Dietrich 1994). Slope stability problems due to rainfall are often encountered in geotechnical engineering, either in tropic regions with frequent rainfall or in arid regions (Tu et al. 2009; Tsaparas et al. 2002; Derbyshire et al. 1995). Even though an otherwise stable slope may fail due to human-induced factors, such as excavation at the toe or loading due to construction, many slopes simply fail due to rainfall infiltration (Ali et al. 2014; Fourie 1996). In the area of Three Gorge Reservoir of China, the frequency of rain-induced landslides accounted for 75% of the total geological disasters since building the reservoir (Li et al. 2011). Hence, the determination of geological mechanisms for the occurrence of the rain-induced gravity erosion is a problem of scientific and societal interest.

The mode of the rain-induced mass failure strongly depends on the initial state of the slope materials, together with the pore water pressure distribution and magnitude of apparent cohesion due to variations in the soil water content (Zhang et al. 2014; Lourenço et al. 2006). Failures may differ in respect of thickness of failed mass, time of failure occurrence, rotational inclination (Au 1998), or be also classified in part by distribution on the ground, in the duration of time that the process acts, by the rate of movement (Wang et al. 2014; Shroder and Bishop 1998). The amount of gravity erosion is pivotal but not readily observable (Xu and Zhao 2014). Site-specific observation is almost impossible due to the uncertainty and non-continuity of gravity erosion (Keefer and Larsen 2007; Benda and Dunne 1997). The volume of individual failure was normally calculated by multiplying the slide area by the thickness of the slide mass (Guzzetti et al. 2009; Haflidason et al. 2005). Nevertheless, the calculated volume involves an amount of tinkering, since the scars caused by the shallow debris flow rapidly heal and are difficult to be detected after a few years (Montgomery and Dietrich 1994). Moreover, erosion volumes caused by water and gravity could not be distinguished with the above approaches.

9.2 Method and Materials

The Loess Plateau is located in the upper and middle reaches of the Yellow River, covering a total area of 624,000 km^2 (Fig. 9.1). The plateau is one of the severest area in the world suffered from gravity erosion (Zhang et al. 2004), and 30–50% of the land is subjected to gravity erosion (Wang et al. 1993). Gravity erosion frequently occurs induced by the rainstorm, because the undulating terrain on the Loess Plateau is characterized by crisscrossing gullies, the vegetation is so sparse, and especially the loess is collapsible due to vertical joints. On the Loess Plateau, a steep bank with the slope of more than 70° in the upper reach of the small watershed is the main source of gravity erosion. Forms of gravity erosion include avalanche, landslide, earth flow, creep and so on (Tang 2004).

Most of the gravity erosions occurred during or after the storms (Ali et al. 2014; Fourie 1996; Montgomery and Dietrich 1994). The specific processes are most easily studied and quantified in a flume using a rainfall simulator under controlled laboratory conditions (Acharya et al. 2011). Here we define and classify the observed mechanisms of rainfall-induced mass failure that occurred on the steep loess slope in the laboratory tests, and then present the latest method to assess the gravity erosion. A topography meter designed by us was used to quantitatively measure the process of gravity erosion.

To classify different failure mechanisms and observe the instability conditions, we conducted a series of gully bank collapse experiments under closely controlled conditions in the Joint Laboratory for Soil Erosion of Dalian University of Technology and Tsinghua University in 2010 and 2012. The landscape simulator was consisted of a rainfall simulator and a slope model covering an area of 3.0 m by 4.0 m (Fig. 9.1). 5 runs of rainfalls each at an amount of 48 mm, were applied in turns on a conceptual landform with the gentle upper slope of 3° and steep lower slope of 70°–80°. An equal period, 12 h or so, was kept after each rainfall to ensure the approximate value of the initial water content. We compacted loess in layers with hands to form the conceptual slope as what we designed before. The 50% diameter of soil particles, d_{50}, was 52.2 μm, and the specific gravity, γ_s, was 2.56.

During and 20 min after the rainfall, slope failure occurrence time, slip mode, type of failure scar, location, and slope failure retrogression behavior were recorded by direct observation and the topography meter. The instrument based on the advanced technology of 3D visualization and stereo video stream display, was designed by the authors themselves, and its performance has been confirmed in the calibration tests and the landslide experiments (Xu et al. 2015). In contrast to the conventional contact observation instruments, the topography meter could quantitatively measure the random mass failures on the steep slope in dynamic environments. The topography meter emitted a group of parallel laser to the slope surface and recorded the dynamic variation of the steep slope relief under rainfall simulation with a video camera. By comparing the slope geometries in the moments before and after the erosion incident on the snapshot images, we could obtain the volume of any individual slide mass.

9.3 Results and Discussion

9.3.1 Types of Observed Mass Failures

The failure style is classified such that it reflects the mechanism responsible for failure movement style, failure block moisture, debris distribution at which the movement occurs. Representative failure modes of all the experiments are illustrated schematically in Fig. 9.2. The phenomenon that soil suddenly topples, fragments and rolls down fully apart from a sloped face is termed avalanche, fall, or collapse, while that soil on the slope slips down as a whole along a certain weak belt is called a landslide. Mudslide, i.e., earth flow or mudflow, is defined here as the failure occurring with distorting shape and involving full saturation.

The maximum volume of the individual failure mass indicates the possible harm on both local constructions and lives. In the experiments shown in Table 9.1, the peak amounts of the individual avalanche, landslide and mudslide are 369.9×10^3, 177.6×10^3 and 24.6×10^3 cm³, respectively. Their average maximum amounts from every experiment are 145.1×10^3, 98.4×10^3, 9.9×10^3 cm³, respectively. Consequently, on the conditions of this experiment, i.e., the steep slope made of homogeneous loess and under intense rainstorms, hazards caused by the avalanche and landslide may be much more serious than that caused by the mudslide. Particularly, the bulk of the individual avalanche was the largest in all types of gravity erosion. As we know, rapid landslides pose lethal threats, whereas slow landslides damage properties but seldom cause fatalities (Iverson et al. 2000). Thus avalanche, which moved quite

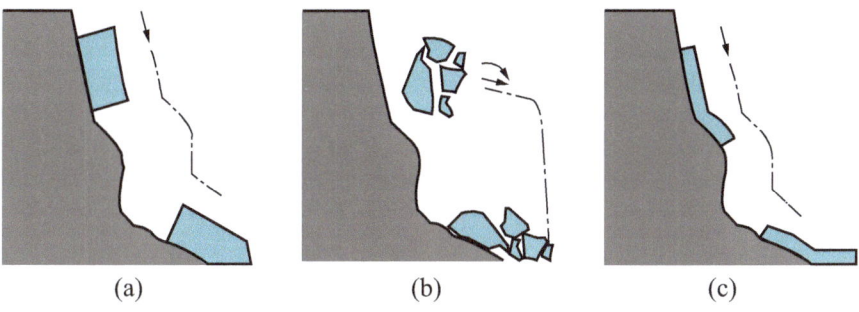

(a)   (b)   (c)

Fig. 9.2 Schematic representation of the gravity erosion types. **a** Landslide; **b** Avalanche; **c** Mudslide. Mudslide was distinguished from the avalanche and the landslide in its high water content and obvious flow performance. The main differences between the avalanche and the landslide are as follows: (1) In the erosion process, the failure block of the avalanche, i.e. avalanche block, was fully separated from a sloped face, whereas that of the landslide slipped down as a whole along a weak belt. (2) After the mass failure, the avalanche block was fragmentized and scattered on the down slope or gully, while the landslide block was a one-piece slope body remarkably characterized with the original lithology

Table 9.1 Volumes of different failure types for an initial landform after 5 rainfalls

Test number	Total volume of a type of initial landform (10^3 cm^3/m)			Maximum volume of the individual failure event (10^3 cm^3)		
	Avalanche	Landslide	Mudslide	Avalanche	Landslide	Mudslide
L1 (S30-1-70d)	172.8	102.4	1.0	41.1	8.2	24.6
L2 (S30-1-80d)	238.9	102.4	1.0	139.5	168.4	2.9
L3 (S30-1.5-70d)	176.6	12.5	7.9	97.2	12.8	12.8
L4 (S30-1.5-80d)	230.1	48.3	2.1	369.9	79.3	6.4
L5 (G60-1-70d)	60.1	104.3	12.2	44.9	177.6	21.7
L6 (G60-1-80d)	72.2	140.9	2.4	60.7	117.7	7.2
L7 (G60-1.5-70d)	161.4	42.5	3.4	224.1	67.5	3.0
L8 (G60-1.5-80d)	94.9	107.1	1.2	183.7	156.0	1.0
Average	150.9	82.5	3.9	145.1	98.4	9.9
Percentage (%)	64	35	2	–	–	–

High-magnitude failure events with the erosion amount more than 500 cm^3 were reckoned in

rapidly downhill, is the most violent gravity erosion and might do great harm to both local constructions and lives.

For any experiment with the same initial landform, the total volumes of the mass failures after 5 rainfalls, which might be transported to the lower reaches by the gully flow, illustrates the potential contribution of gravity erosion to the soil loss from the catchment. Avalanche and landslide were the most frequently observed failures, whereas mudslide was observed only on a few occasions and its size was relatively small during the experiments. As shown in Table 9.1, the average value of the total

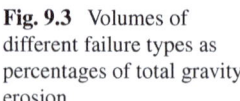

Fig. 9.3 Volumes of different failure types as percentages of total gravity erosion

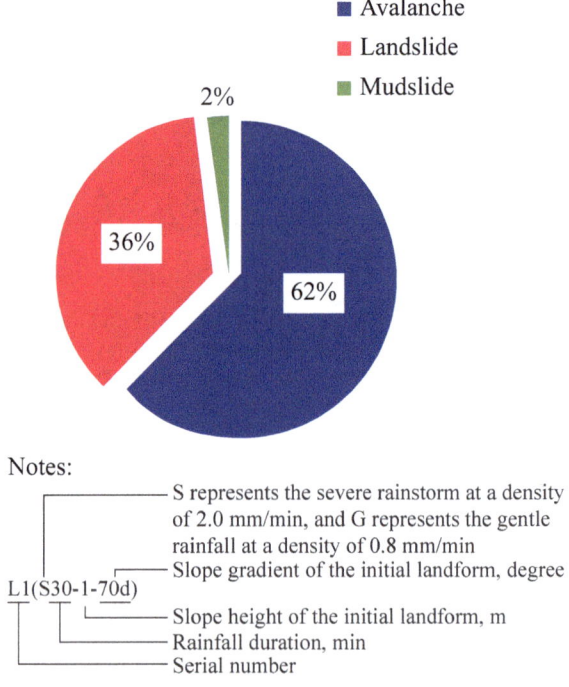

Notes:

L1(S30-1-70d)

S represents the severe rainstorm at a density of 2.0 mm/min, and G represents the gentle rainfall at a density of 0.8 mm/min

Slope gradient of the initial landform, degree

Slope height of the initial landform, m

Rainfall duration, min

Serial number

volume of avalanche, landslide and mudslide for each experiment were 150.9×10^3, 82.5×10^3, 3.9×10^3 cm^3/m, respectively. That's to say, the amount of avalanche, landslide and mudslide accounted for 62, 36 and 2% of the total gravity erosion in a rainfall experiment of the model test (Fig. 9.3). Avalanche and landslide, especially the former, played a crucial role as the landform was made with loess by hand patting. A field investigation in the Northern Weihe River o f he Loess Plateau showed that, the amount of avalanche and landslide accounted for 96% of the total gravity erosion (Liu et al. 1990). The statistical data are qualitatively in good agreement with the observations in Table 9.1.

9.3.2 Triggers of the Gravity Erosion

Several factors, including the characteristics of the rainfall, the saturated hydraulic conductivity of soil, the slope geometry, the initial conditions and the boundary conditions, affect the stability of a slope subjected to rainfall infiltration (Ali et al. 2014). In practice, people usually improve the landform with the vegetation or structural engineering to alleviate gravity erosion. To assess the effects of the initial landform geometry on the gravity erosion, experiments in this study are divided into 4 groups. Each had the same slope height or gradient. For all groups, the maximum of the

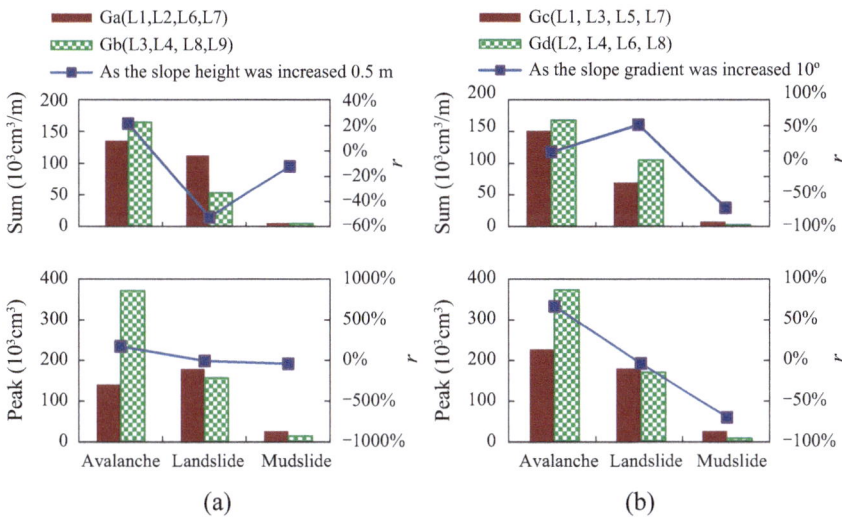

Fig. 9.4 Changes of the gravity erosion impacted by the landform geometry. The difference of erosion volumes before and after the initial landform changed is $r = (b - a)/a \times 100\%$, where r is the augment of the total gravity erosion or the maximum individual failure; a is the erosion amount before the impact factor changed; and b is the erosion amount after the impact factor changed

individual failure masses and the average values of the total erosion of every experiment are compared, as shown in Fig. 9.4. An advantage of the above method is the influence caused by randomness of the gravity is handily overcome.

The potential energy of a collapsed mass, which depends on the block's volume and height, has a direct impact on the erosion amount. While other conditions were fixed but the slope height was increased from 1.0 to 1.5 m, the total amount of avalanche increment was 22% and the maximum volume of individual avalanche was augmented by 165%. In contrast, landslide and mudflow are soil slides under gravity along the slope. Their potential energies are weakened by the friction of slope, so that the influence of the slope height is relatively small. The slope's steepness also has an important influence on gravity erosion. When other conditions were fixed and the slope gradient was increased from 70° to 80°, the volume of avalanche and landslide increased, of which the total amount of landslide was increased by 52%, and the maximum amount of individual avalanche was increased by 65%.

9.3.3 Prevention and Control Measures

The control methods may be different for various types of gravity erosion. Changes in vegetation cover often result in modified landslide behavior (Glade 2003). While some of the hydrological and geological changes are difficult to predict, this hazard can be also mitigated by appropriate geotechnical control or by regulation of human activities, proper engineering works, maintenance (Wei et al. 2006; Au 1998). Installing retaining facilities, and draining surface and underground water from the sliding area, are all effective measures in the control of avalanche, landslide and earth flow. Additionally, cutting the steep slopes to make them gentler is usually adopted in the area prone to avalanches. Stabilization measures for landslide also include: (1) excavating and redistributing sliding mass, and (2) coarsening the slip band. Besides the landslide and avalanche control methods, planting also plays a more important role in controlling mudslide, while drainage works, e.g., diversion dike and chute, are particularly used to control the mudflow.

On the Loess Plateau of China, gravity erosion is remarkably controlled by the check-dams. The base level of erosion will be raised while the dam farmland is formed (Xu et al. 2004, 2006). Thus the height and mean gradient of the bank are decreased (Fig. 9.4). Moreover, toes of the slopes are also compacted and strengthened, the longitudinal gradient of the gully is decreased, also the scour and transport capacity of the flow in the gully is reduced. Hence avalanche, landslide, and earth flow are all alleviated as the check dam is built.

9.4 Conclusions

We conclude that as heavy rainfall applied on the steep slope compacted by hand with loess, soil loss caused by the avalanche and landslide is much more serious than that caused by the mudslide, and especially the avalanche is the most violent gravity erosion which might do great harm to the local transportations and lives. Furthermore, the slope height and gradient had a remarkable impact on the erosion amount.

References

Acharya G, Cochrane T, Davies T, et al. 2011. Quantifying and modeling post-failure sediment yields from laboratory-scale soil erosion and shallow landslide experiments with silty loess. Geomorphology, 129(1–2): 49–58.

Ali A, Huang J S, Lyamin A V, et al. 2014. Boundary effects of rainfall-induced landslides. Computers and Geotechnics, 61: 341–354.

Au S W C. 1998. Rain-induced slope instability in Hong Kong. Engineering Geology, 51(1): 1–36.

Benda L, Dunne T. 1997. Stochastic forcing of sediment supply to channel networks from landsliding and debris flow. Water Resources Research, 33(12): 2849–2863.

Derbyshire E, Van Asch T, Billard A, et al. 1995. Modelling the erosional susceptibility of landslide catchments in thick loess: Chinese variations on a theme by Jan de Ploey. Catena, 25(1–4): 315–331.

Fourie A B. 1996. Predicting rainfall-induced slope instability. Proceedings of the Institution of Civil Geotechnical Engineering, 119(4): 211–218.

Glade T. 2003. Landslide occurrence as a response to land use change: a review of evidence from New Zealand. Catena, 51(3–4): 297–314.

Guzzetti F, Ardizzone F, Cardinali M, et al. 2009. Landslide volumes and landslide mobilization rates in Umbria, central Italy. Earth and Planetary Science Letters, 279(3–4): 222–229.

Haflidason H, Lien R, Sejrup H P, et al. 2005. The dating and morphometry of the Storegga Slide. Marine and Petroleum Geology, 22(1–2): 123-136.

Iverson R M, Reid M E, Iverson N R, et al. 2000. Acute Sensitivity of landslide rates to initial soil porosity. Science, 290(5491): 513–516.

Keefer D K, Larsen M C. 2007. Assessing landslide hazards. Science, 316(5828): 1136–1138.

Li H Z, Yang Z S, Wang T R, et al. 2011. Recognition of landslide and case studies. Wuhan: Wuhan University of Technology Press: 19–43 (in Chinese).

Liu B, Zhai M, Wu F. 1990. Gully erosion on the North Weihe Plateau. Memoir of NISWC. Mem NISWC, Academia Sinica and Ministry of Water Conservancy, 12: 25–33 (in Chinese).

Lourenço S D N, Sassa K, Fukuoka H. 2006. Failure process and hydrologic response of a two layer physical model: Implications for rainfall-induced landslides. Geomorphology, 73(1): 115–130.

Montgomery D R, Dietrich W E. 1994. A physically based model for the topographic control on shallow landsliding. Water Resources Research, 30(4): 1153–1171.

Peruccacci S, Brunetti M T, Luciani S, et al. 2012. Lithological and seasonal control on rainfall thresholds for the possible initiation of landslides in central Italy. Geomorphology, 139(4): 79–90.

Salciarini D, Godt J W, Savage W Z, et al. 2006. Modeling regional initiation of rainfall-induced shallow landslides in the eastern Umbria Region of central Italy. Landslides, 3(3): 181–194.

Shroder Jr J F, Bishop M P. 1998. Mass movement in the Himalaya: new insights and research directions. Geomorphology, 26(1–3): 13–35.

Tang K L. 2004. Soil and water conservation in China. Beijing: Science Press: 100–104 (in Chinese).

Tsai T L, Yang J C. 2006. Modeling of rainfall-triggered shallow landslide. Environmental Geology, 50(4): 525–534.

Tsaparas I, Rahardjo H, Toll D G, et al. 2002. Controlling parameters for rainfall-induced landslides. Computers and Geotechnics, 29(1): 1–27.

Tu X B, Kwong A K L, Dai F C, et al. 2009. Field monitoring of rainfall infiltration in a loess slope and analysis of failure mechanism of rainfall-induced landslides. Engineering Geology, 105(1–2): 134–150.

Wang D F, Zhao X Y, Ma H L, et al. 1993. An investigation on the gravity erosion on the loess area. Soil and Water Conservation in China, (12): 26–28 (in Chinese).

Wang F X, Zhang C, Peng Y, et al. 2014. Diurnal temperature range variation and its causes in a semiarid region from 1957 to 2006. International Journal of Climatology, 34(2): 343–354.

Wei J B, Xiao D N, Xie F J. 2006. Evaluation and regulation principles for the effects of human activities on ecology and environment. Progress in Geography, 25(2): 36–45 (in Chinese).

Xu X Z, Zhao C. 2014. A laboratory study for gravity erosion of the steep loess slopes under intense rainfall. Proceeding of 11th International Conference on Hydroscience and Engineering, Hamburg: 709–715.

Xu X Z, Zhang H W, Zhang O Y. 2004. Development of check-dam systems in gullies on the Loess Plateau, China. Environmental Science and Policy, 7(2): 79–86.

Xu X Z, Zhang H W, Wang G Q, et al. 2006. A laboratory study on the relative stability of the check-dam system in the Loess Plateau, China. Land Degradation and Development, 17(6): 629–644.

Xu X Z, Zhang H W, Wang W L, et al. Quantitative monitoring of gravity erosion using a novel 3D surface measuring technique: validation and case study. Natural Hazards, 75: 1927–1939.

Zhang C S, Zhang Y C, Zhang L H. 2004. Danger assessment of collapses, landslides and debris flows of geological hazards in China. Journal of Geomechanics, 10(1): 27–32 (in Chinese).

Zhang C, Wang D G, Wang G L, et al. 2014. Regional differences in hydrological response to canopy interception schemes in a land surface model. Hydrological Processes, 28(4): 2499–2508.

Chapter 10
A Sensitivity Analysis on the Gravity Erosion on the Steep Loess Slope

Abstract Gravity erosion is a dominant geomorphic process on the widespread steep loess slopes, yet it is not well understood due to the complexity of failure occurrence and behavior. This chapter conducted a series of laboratory experiments to test the stability of different slope geometries and rainfalls and then performed a sensitivity analysis. The following three types of gravity erosion were observed: landslide, avalanche, and mudslide. In an event of rainfall, various types of gravity erosion might emerge in the same period, and mass failures with the same mode and similar size often adjacently appeared. Climate-driven factors and topography triggers had prominent influences on gravity erosion. Whether for the total amount or the peak amount in an experiment, the largest sensitivity parameter on both landslides and mudslides was that of rainfall duration. The experimental results provide an insight into the pre-failure mechanisms and processes of steep loess slopes.

Keywords Gravity erosion · Laboratory test · Sensitivity analysis · Rainfall · Landform

10.1 Assessment of Soil Erosion Sensitivity

Assessment of soil erosion sensitivity is defined as the possibility of soil erosion occurrence and identification of areas susceptible to soil erosion when only considering natural factors (Zhang et al. 2013). The main task in landslide susceptibility assessment is to find out how the causal factors influence the occurrence of landslides (Melchiorre et al. 2011). The mode of rain-induced mass failure strongly depends on the initial state of the slope materials, together with the pore water pressure distribution and magnitude of apparent cohesion due to variations in the soil water content (Zhang et al. 2014; Lourenço et al. 2006). Intense rainfall, soils that are largely non-cohesive as they become saturated, steep terrain, and intense development are considered to be the major causes of the failures. Loss of pore-water suction, erosion, and pore-water pressure build-up at shallow depths are the most common ways through which rainwater affects slope stability, as short-burst rainstorms are common. The scale of a failure event depends on the intensity, area, position, and duration of the triggering rainstorm, whereas the antecedent rainfall has relatively

© Science Press and Springer Nature Singapore Pte Ltd. 2020
X. Xu et al., *Experimental Erosion*,
https://doi.org/10.1007/978-981-15-3801-8_10

little influence (Au 1998). The erosional history and the consequent morphology are also much more important except for the trimming induced by occasional very large run-off events (Thornes and Alcántara-Ayala 1998).

Because gravity erosion is affected and constrained by so many factors, its quantification is complicated and difficult to achieve. Furthermore, gravity erosion is a stochastic, non-continuous process, and usually occurs as a combination of soil transportation with sheet flow and mass failure on the steep slope (Keefer and Larsen 2007; Benda and Dunne 1997). Although, the process is readily observed on natural hill slopes, quantifying it in a natural environment is significantly challenging given the extended timeframe between the occurrence of the process, and variability in rainfall, soils, and other factors (Acharya et al. 2011). Site-specific and real-time measurement is almost impossible due to the uncertainty and non-continuity of gravity erosion. Hence the volume of individual failure was normally calculated by multiplying the slide area by the thickness of the slide mass after the rainfall events (Guzzetti et al. 2009; Haflidason et al. 2005). Nevertheless, the calculated volume involves an amount of tinkering, for shallow debris-flow scars rapidly heal and are difficult to detect after as few as years (Montgomery and Dietrich 1994). Moreover, erosion volumes caused by water and gravity could not be distinguished from the above calculation approaches. Landslide activity maps represent a short-cut in the assessment of mass movement hazards (Parise and Wasowski 1999). While valuable, these inventory maps usually do not provide information on the timing of the events, making it difficult to correlate landslide occurrences with specific triggering events (Kirschbaum et al. 2010).

The specific processes of rain-induced mass failures are most easily studied and quantified in a flume using a rainfall simulator under controlled laboratory conditions (Acharya et al. 2011). Here, we employed a topography meter designed by us to quantitatively measure the process of gravity erosion, and we utilized the increase-rate-analysis method to analyze the sensitivity of gravity erosion. The experimental activity was focused on processes related to gravity and to the interaction between rainfall and topography.

10.2 Gravity Erosion on the Loess Plateau

Areas of the Loess Plateau, especially the Loess Hill Ravine Region and the Loess Mesa Ravine Region, are severely affected by gravity erosion (Figs. 6.1 and 10.1). All types of mass failure are abundant in the area, and locally cover 30–50% of the land (Wang et al. 1993). In the area, rainstorm-induced gravity erosion frequently occurs, because the undulating terrain on the Loess Plateau is characterized by crisscrossing gullies, the vegetation is so sparse, and especially the loess is collapsible and in vertical joints. On the Loess Plateau, a steep bank with the slope of more than 70° in the upper reaches of the small watershed is the main source of gravity erosion. Forms of gravity erosion on the Loess Plateau include avalanche, landslide, earth flow, and creep (Tang 2004).

(a)   (b)

Fig. 10.1 Study area

10.3 Experimental Methods

To classify different failure mechanisms and observe conditions of instability, we conducted a series of gully bank collapse experiments under closely controlled conditions in 2010 and 2012 in the Joint Laboratory for Soil Erosion of Dalian University of Technology and Tsinghua University located in Beijing, China. The landscape simulator consisted of a rainfall simulator and a slope model covering an area of 3.0 m by 3.0 m (Fig. 10.2). Five runs of rainfall were applied in turn on a conceptual landform with a gentle upper slope of 3° and steep lower slope of 70°–80°. An equal period, 12 h or so, was kept after each rainfall to ensure the approximate value of initial water content. The conceptual slope was made with loess by hand patting. The 50% diameter of soil particles, d_{50}, was 52.2 μm, and the specific gravity, γ_s, was 2.56. The physical properties of the model soil was similar to that of the Loess Plateau; that is, distribution of the grain size is close that in Shanxi, Gansu, and Shannxi (Xu et al. 2009). A summary of the tests carried out by us is reported in Table 10.1.

In this experimental study, the failure style was defined by direct eye observation of the process of soil deformation, and the volume of failure mass was calculated according to the video of the topography meter. Both during and 20 min after the rainfall, slope failure occurrence time, slip mode, type of failure scar, location, and slope failure retrogression behavior were recorded by direct observation and the Topography meter (Fig. 10.2). In contrast to the conventional contact observation instruments, the topography meter could quantitatively measure the random mass failures on the steep slope in dynamic environments. The topography meter emitted a group of parallel lasers to the slope surface and recorded the dynamic variation of the steep slope under rainfall simulation with a video camera. Then the operator could transform the plane figures into 3D graphs to compute the shape of the target surface. By comparing the slope geometries in the moments before and after the erosion incident on the snapshot images, we could obtain the soil erosion data, including the volume of any individual slide masses. The instrument was invented

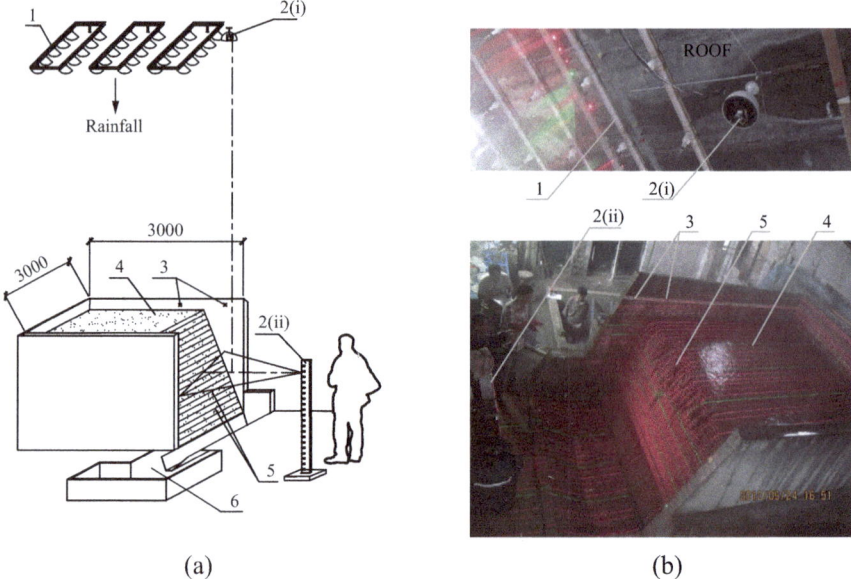

(a)   (b)

Fig. 10.2 Landscape simulator in which rainfall simulation experiments were conducted. All units are in millimeters. **a** Blue print of the topography meter measurement system; **b** Pictures of the experimental site. Keys: 1. Rainfall simulator; 2. Topography meter (i Camera with a collimator and ii Laser source); 3. Positioning marks; 4. Model slope; 5. Equidistant horizontal projections; 6. Receiving pool

Table 10.1 Testing program

Test number	Lower slope configuration		Rainfall		
	Height (m)	Gradient (°)	Intensity (mm/min)	Duration (min)	Runs
L1	1.0	70	2.0	30	5
L2	1.0	80	2.0	30	5
L3	1.5	70	2.0	30	5
L4	1.5	80	2.0	30	5
L5	1.0	70	0.8	60	5
L6	1.0	80	0.8	60	5
L7	1.5	70	0.8	60	5
L8	1.5	80	0.8	60	5
L9	1.0	70	0.8	30	5
L10	1.0	80	0.8	30	5

by the authors themselves, and its performance was confirmed in the calibration tests and the landslide experiments (Xu et al. 2015a, b).

Gravitational erosion involved both large-scale mass wasting and smaller-scale erosion. The size of each mass failure was calculated and classified, and then the total amount of all failure masses g_t and the peak value of individual erosion events during a rainfall event g_p were obtained. All failure masses with a volume of more than 500 cm^3 were considered in the experimental study. To assess the effects of the initial landform geometry on the gravity erosion, we divided experiments into the following eight experimental groups, each of which had the same slope height or gradient:

G1 (Experiments L5–8) versus G2 (Experiments L1–4). Rainfall intensity in the former experimental group was 0.8 mm/min, while the later was 2.0 mm/min.
G3 (Experiments L9–10) versus G4 (Experiments L5–6). Rainfall duration in the former experimental group was 30 min, while the later was 60 min.
G5 (Experiments L1, 3, 5 and 7) versus G6 (Experiments L2, 4, 6 and 8). Slope gradient of the initial lower slope in the former experimental group was 70°, while the later was 80°.
G7 (Experiments L1, 2, 5 and 6) versus G8 (Experiments L3, 4, 7 and 8). Slope height of the initial lower slope in the former experimental group was 1.0 m, while the latter was 1.5 m.

For all groups, the maximum of the individual failure masses and the average value of the total erosion of every experiment were compared. An advantage of the above method is that influences caused by the randomness of gravity could be handily overcome.

Then we used the increase-rate-analysis method to evaluate variations in the gravity erosion with respect to changes in other causal factors such as rainfall intensity and duration, and slope gradient and height. Our sensitivity analysis focuses on g_t and g_p. The increase ratio of gravity erosion R_g (%) is shown as follows:

$$R_g = (\bar{g}_2 - \bar{g}_1)/\bar{g}_1 \tag{10.1}$$

where $\bar{g}_1$ is the average value of g_t or the maximum value of g_p before the triggering element was changed in an experimental group, cm^3, and $\bar{g}_2$ is that after the triggering element was changed, cm^3. An increased ratio of the above output in percent will be calculated with the growth rate of a parameter while other parameters are fixed in an experimental group:

$$S = R_g/R_t \tag{10.2}$$

where S is the sensitivity parameter to analyze the sensibility of the failure volume to the triggering elements; $R_t = (t_2 - t_1)/t_1$ is the increased ratio of the triggering element, namely rainfall or landform, in which t_1 is the value before being changed in an experimental group, and t_2 is that after being changed. R_t is conveyed in percent. This approach allows for a qualitative investigation of the effect of conditioning factors.

10.4   Results and Discussion

10.4.1   Occurrence and Behavior of the Gravity Erosion

The failure style is classified such that it reflects the mechanism responsible for failure movement style, failure block moisture, and debris distribution at the location where the movement occurs. All three types of gravity erosion have been observed: landslide, avalanche, and mudslide. A summary of the evolution of slope profiles during events of gravity erosion is given in Figs. 10.3, 10.4 and 10.5 with reference to the images from the digital video camera. The phenomenon that soil suddenly topples, fragments, and rolls down fully apart from a sloped face is termed avalanche, fall, or collapse; while rock soil on the slope that slips down as a whole along a certain surface of rupture is called a landslide. Avalanche is a violent mass failure occurring at steep hillsides. Once released, it may move quite rapidly downhill. The avalanche block is fragmentized, and fully separated from the mother land. Landslide is remarkably characterized with the original properties of the slide block (Li et al. 2011). Generally, the block of landslide integrally moves along the distinct fracture plane—in some cases favored by water, sometimes advancing very quickly, others creeping slowly. Mudslide, i.e., earth flow or mudflow, is defined here as a failure occurring with the

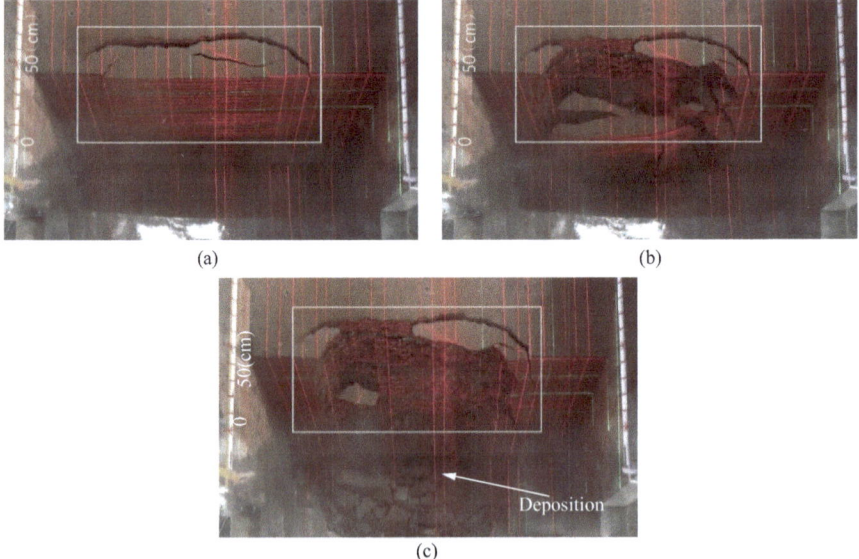

Fig. 10.3   An avalanche during the first rainfall of Test L5. **a** Elapsed time, ET in short, is 58′54″ (58′54″ after the rainfall started). Crevices were creating and expanding which indicated a landslide was coming. **b** ET was 59′01″. The failure block was toppling down, fully separated from the mother land. **c** ET was 59′05″. The failure block was fragmentized and scattered in the gully

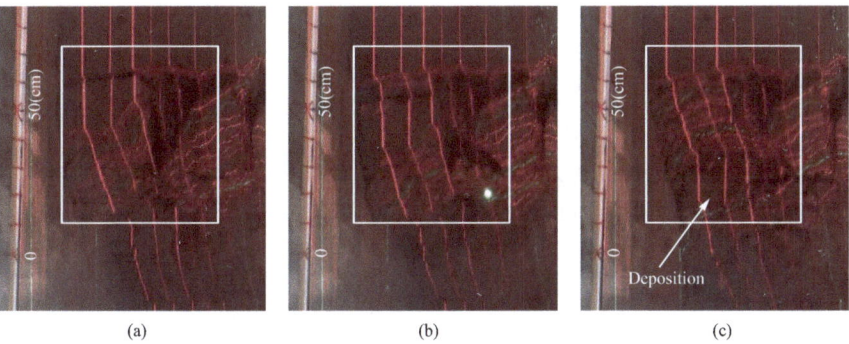

(a)   (b)   (c)

Fig. 10.4 A landslide during the fourth rainfall of Test L6. **a** ET was 37′21″. A landslide started. **b** ET was 37′32″. The landslide block was slipping down as a whole close to the mother land. **c** ET was 37′51″. The landslide block was a one-piece slope body remarkably characterized with the original lithology as the event ended

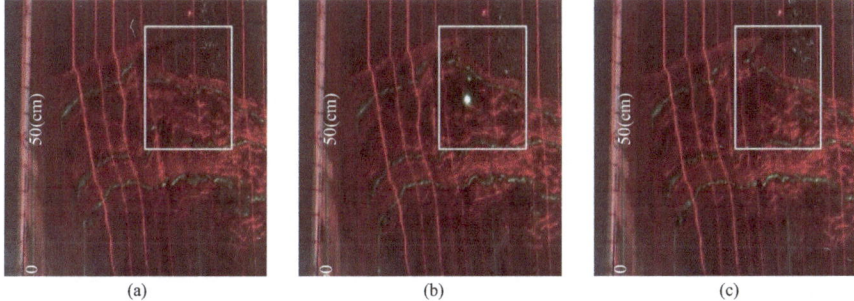

(a)   (b)   (c)

Fig. 10.5 A mudslide during the fourth rainfall of Test L6. **a** ET was 58′55″. Landform before a mudslide. **b** ET was 59′18″. A mudflow was initiated at the top-left corner of the white frame. **c** ET was 59′48″. The mud flow with high water content was in close proximity to liquid form

distorted shape of the slope and involving full saturation. It causes full or partial soil liquefaction due to high pore-pressures and may trigger highly mobile soil loss.

In the event of rainfall, various types of gravity erosion may emerge in the same period. Landslide and avalanche grow out of similar geological tectonic environments and lithology structure conditions, and are triggered by the same factors; so landslides in an area are often accompanied by avalanches. On the other hand, mass failures with the same mode and similar size often appear adjacently in the same run of rainfall. For example, 38 min after the start of the second rainfall of Experiment L8, a landslide with a volume of 6.75×10^4 cm^3 happened; after 18 min (elapsed time was 56 min), another landslide with the magnitude of 6.3×10^4 cm^3 occurred. Forty-eight minutes after the start of the third rainfall of Experiment L8, a landslide with a volume of 4.16×10^4 cm^3 happened; 11 min later (elapsed time was 59 min), another landslide with a magnitude of 3.83×10^4 cm^3 occurred.

Some mass movement processes act very slowly, while others occur suddenly. In present laboratory experiments, an interesting phenomenon, termed compound landslide, was observed: a group of mass failures including avalanche, mudslide, or landslide might happen on a large, slowly slipping block. The process could last for several minutes, or more than 10 min during a 1 h rainfall event. To distinguish the compound landslide from other individual gravity erosion events has practical significance. When we calculate the total amount of gravity erosion in a rainfall event, we could calculate the volume of the compound landslide with the screenshots in the moments before and after the longstanding landslide and need not care about the incidents that happened on the sliding block. Surely, deviation resulted from the hydraulic erosion still existed, but the error was so small compared to the large landslide that it could be ignored. Figure 10.6 shows an event of compound mass failure on the steep loess slope in the fourth rainfall of Test L5. When a large block was sliding, two avalanches in relatively small bulk happened on the block. Because the avalanches occurred on the slip mass during the compound landslide, the erosion

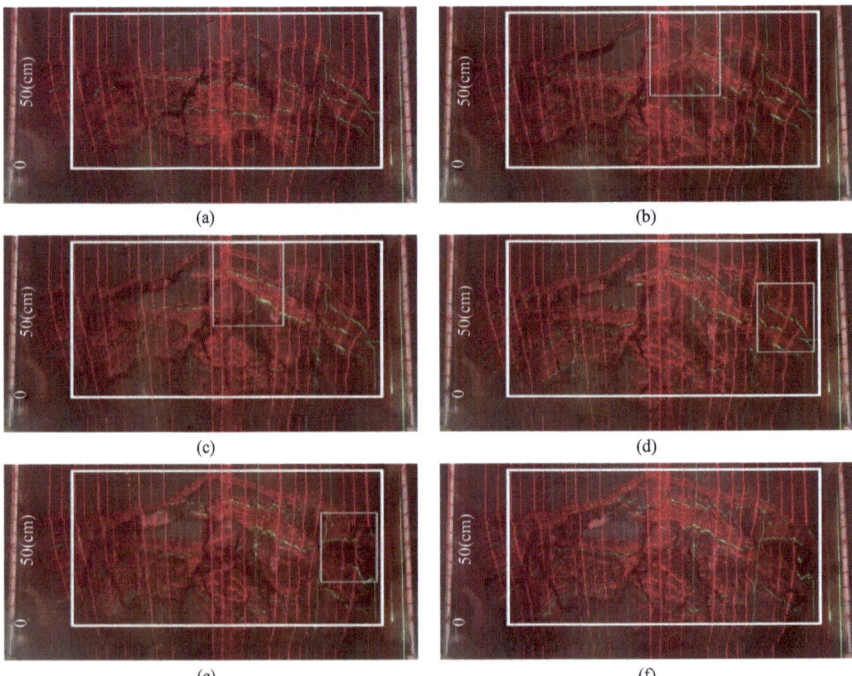

Fig. 10.6 An event of compound mass failure during the fourth rainfall of Test L5. Because the avalanches occurred on the slip mass in the period of the compound landslide, the erosion volume of the compound landslide was the difference of the slope volume in the white frame between **a** and **f**. **a** ET was 39′26″. This is the terrain before a compound landslide. **b** ET was 40′32″ and **d** ET was 41′09″. Landforms in the small white frame rectangles are those before avalanches happened. **c** ET was 40′50″ and **e** ET was 41′43″. Landforms in the small white frame rectangles are those after avalanches happened. **f** ET was 42′11″. Here is the failure scar caused by the compound landslide

volume of the compound landslide was the difference of the slope volume in the white frame among Figs. 10.6a, f.

10.4.2 Triggers: Rainfall Intensity and Duration

Changes in rainfall and terrain severely influence sediment dynamics, e.g., erosion and retention processes (Sánchez-Canales et al. 2015; Ali et al. 2014). The erosional history and the consequent morphology are also much more important than the trimming induced by occasional very large run-off events (Thornes and Alcántara-Ayala 1998). For any experiment with the same initial landform, the total volume of the mass failures after five rainfalls, which might be transported to the lower reaches by the gully flow, illustrated the potential contribution of gravity erosion to the soil loss from the catchment. On the other hand, the maximum volume of the individual failure masses indicated possible harm to the local building construction or human lives. For all groups, the average value of total erosion and the maximum of individual failure masses of every experiment are listed in Tables 10.2 and 10.3. Here, WGE, Whole Gravity Erosion, means the sum of avalanche, landslide, and mudslide.

Rainwater infiltrating to soil might decrease the anti-shearing strength of the failure surface, and consequently induce gravity erosion on the steep slope (Li et al. 2011). Rainfall intensity and duration are the most immediate impact factors on gravity erosion. Inspection of Fig. 10.7 reveals that the total volume of the avalanches was enlarged by 111% and the maximum volume of the individual avalanches was augmented by 65% if the rainfall intensity was increased from 0.8 to 2.0 mm/min, namely: a 150% increase in rainfall intensity. At the same time, the amounts of landslide and mudslide were only a little reduced. Hence a rainstorm with high

Table 10.2 Total volumes of different failure types for an initial landform after five rainfalls

Experimental group	Average of a type of initial landform (10^3 cm^3/m)			
	Avalanche	Landslide	Mudslide	WGE
G1 (L5–L8)	97.1	98.7	4.8	200.6
G2 (L1–L4)	204.6	66.4	3.0	274.0
G3 (L9–L10)	52.4	4.7	0.4	57.5
G4 (L5–L6)	66.1	122.6	7.3	196.0
G5 (L1, L3, L5, L7)	142.7	65.4	6.1	214.3
G6 (L2, L4, L6, L8)	159.0	99.7	1.7	260.3
G7 (L1, L2, L5, L6)	136.0	112.5	4.1	252.6
G8 (L3, L4, L7, L8)	165.7	52.6	3.6	222.0

High-magnitude failure events with the erosion amount more than 500 cm^3 were analyzed
Notes WGE, namely the whole gravity erosion, means a sum of avalanche, landslide and mudslide

Table 10.3   The maximum volume of a type of mass failure for an initial landform after five rainfalls

Experimental group	Peak of the individual failure events (10^3 cm^3)			
	Avalanche	Landslide	Mudslide	WGE
G1 (L5–L8)	224.1	177.6	21.7	224.1
G2 (L1–L4)	369.9	168.4	24.6	369.9
G3 (L9–L10)	75.2	27.1	1.3	75.2
G4 (L5–L6)	60.7	177.6	21.7	177.6
G5 (L1, L3, L5, L7)	224.1	177.6	24.6	224.1
G6 (L2, L4, L6, L8)	369.9	168.4	7.2	369.9
G7 (L1, L2, L5, L6)	139.5	177.6	24.6	177.6
G8 (L3, L4, L7, L8)	369.9	156.0	12.8	369.9

High-magnitude failure events with the erosion amount more than 500 cm^3 were analyzed

Fig. 10.7   Increment of the mass failure as the rainfall intensity was increased by 150%

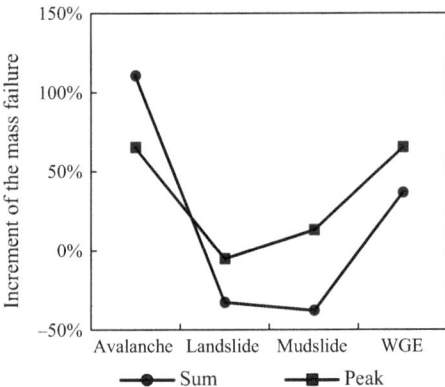

intensity and long period could give rise to avalanches with a relatively larger range and bulk.

Figure 10.8 portrays increment of the mass failure as the rainfall duration increased. If the rainfall intensity was kept constant at a low intensity, 0.8 mm/min, while the duration increased from 30 to 60 min—namely, a 100% increase in rainfall duration—the volume of avalanche was not relatively obviously enlarged, but total volumes of the landslide and mudslide were augmented to 24.9 and 19.5 times, and the maximum volumes of individual landslide and mudslide were augmented 5.5 and 15.6 times, respectively. In conclusion, the loess slope under light rain is more easily subjected to landslide and mudflow as rainfall increases in duration.

In the experiments, a clear relationship has been found between the failure style and the rainfall mode, i.e., intensity or duration. Nevertheless, the result is so different from that in other geographic regions described b y Lourenço et al. (2006), where no clear relation was found between the pore water pressure and the failure mode.

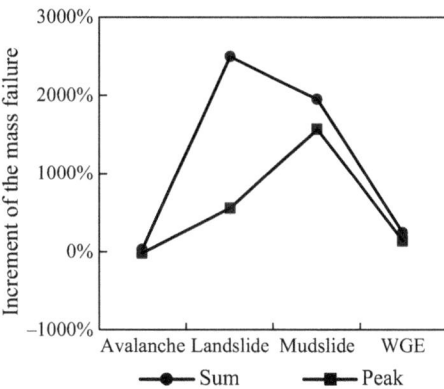

Fig. 10.8 Increment of the mass failures as the rainfall duration was increased by 100%

10.4.3 Triggers: Slope Height and Gradient

Topography is another significant factor associated with soil erosion. An increase of slope gradient will result in enlarged gravity erosion. The scatter diagram in Fig. 10.9 shows an increment of the mass failure as the slope gradient was increased. When other conditions were fixed but the slope gradient increased from 70° to 80°—i.e. a 14% increase in slope gradient—the total amount of landslide was increased by 52%, and the maximum amount of individual avalanche was increased by 65%. Nevertheless, both the total and maximum amounts of the mudslide were decreased by more than 70%. In fact, as the slope gradient grew, the total amount of WGE was increased by 21%, by virtue of the relatively large erosion bulks of the avalanche and landslide.

Slope height also has an important influence on gravity erosion. Avalanche is a high-speed movement of the soil mass torn instantly from its parent. The potential energy of a collapsed mass, which depends on its volume and height, has a direct impact on the erosion amount. While other conditions were fixed but the slope height

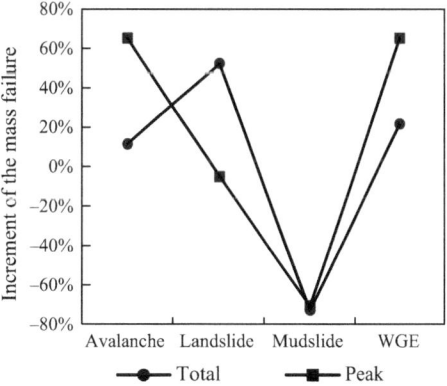

Fig. 10.9 Increment of the mass failure as the slope gradient was increased by 14%

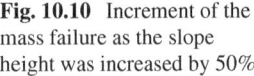

Fig. 10.10 Increment of the
mass failure as the slope
height was increased by 50%

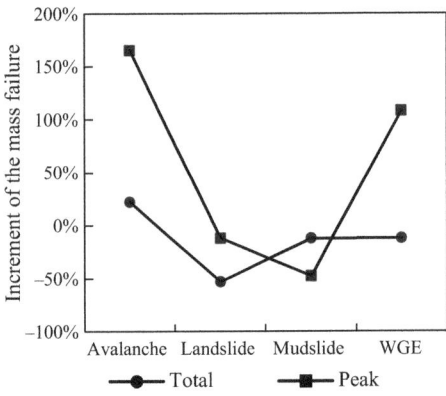

increased from 1.0 to 1.5 m—i.e. a 50% increase in slope height—the total amount of
avalanche was increased by 22%, and the maximum volume of individual avalanche
was augmented by 165%, as shown in Fig. 10.10. In contrast, landslide and mudflow
are soil slides drawn by gravity along the slope. Their potential energies are weakened
by the friction of slope, so that the influence of the slope height is relatively small. To
our surprise in the experiments, a small decrease occurred on the amounts of landslide
and mudslide when the slope height increased from 1.0 to 1.5 m. The reason may be
the randomness of the gravity erosion or the deviation of the experiment.

10.4.4   Causes of Different Failures: A Concise Discussion

Factors that induce gravity erosion include the following two main categories (Liu
et al. 2013; Wu and Sidle 1995): (1) internal factors that have decisive effects on
landslides, i.e. geology, geomorphology, soil property, vegetation cover, flow distribution and fractures, and (2) external factors that trigger landslides suddenly, such
as rainfall, earthquake and flood. Mass failures are common on the Loess Plateau
due to the presence of macropores, well-developed vertical jointing, and susceptibility to water infiltration (Zhang and Liu 2010; Zhang et al. 2009). Loess has
typical landforms like vertical joints and loose textures, as well as special physical
and mechanical properties, such as low water content, strong structural strength, and
weaker water resistance (Li et al. 2013). Most mass failures are triggered by slope
cutting and heavy rainfall (Zhuang and Peng 2014). Enhanced by human activities,
precipitation infiltrates into the interior of the loess formations along structural joints
and openings caused by weathering (Zhang et al. 2015). Steep-cut slopes encourage
the concentration of shear stress at the foot of the slope and tension stress at the top,
leading to the formation of cracks in the inner slope, and followed by slope failure.

It could be seen that on the sides of the platforms, ridges and domes are the steep
slopes which could easily slide in favorable to soil failure, such as infiltration of rain

water, irrigation water, and earthquake effects. Generally, loess landslides occur on slopes steeper than 35° and higher than 40 m, and the sliding is more likely to occur on the concave slopes (Li et al. 2013). An investigation on the Loess Plateau by Cao (1981) illustrates that landslides generally happened on a slope with a gradient of 35°–55°, and most of the avalanches occurred on scarps steeper than 55°. Landslides were common in the middle and lower reaches of the gully, whereas avalanches were frequent on the banks of the gully or of the river, especially on the gully head. Investigations in the northern Shaanxi Province (Li et al. 2013) shows that 80% of loess landslides occurred in slopes with dip angles exceeding 35°, and failure style was normally collapse-sliding or purely collapsed as the slope angle became greater than 50°. On the other hand, increments of different failures due to landform and rainfall have been described in parts 10.3.2 and 10.3.3 of this paper.

10.4.5 Sensitivity Analysis

Sensitivity analyses were conducted to evaluate the conditions required for the initiation of storm-induced gravity erosion. Our analysis identified the parameters related to the natural environment as the most influential for failure occurrence and behavior. The results are summarized in Fig. 10.11a, b. It is clear that with the increase of factors of rainfall duration-intensity and slope height-gradient, in most cases both the total and peak amount of gravity erosion during a rainfall event were enlarged. For the total amount in an experiment, the sensitivity parameters of rainfall duration on the landslide, mudslide, and WGE were 24.9, 19.5 and 2.4, respectively, and the sensitivity parameter of rainfall intensity on the avalanche was 2.2. For the peak amount in an experiment, the sensitivity parameters of rainfall duration on the

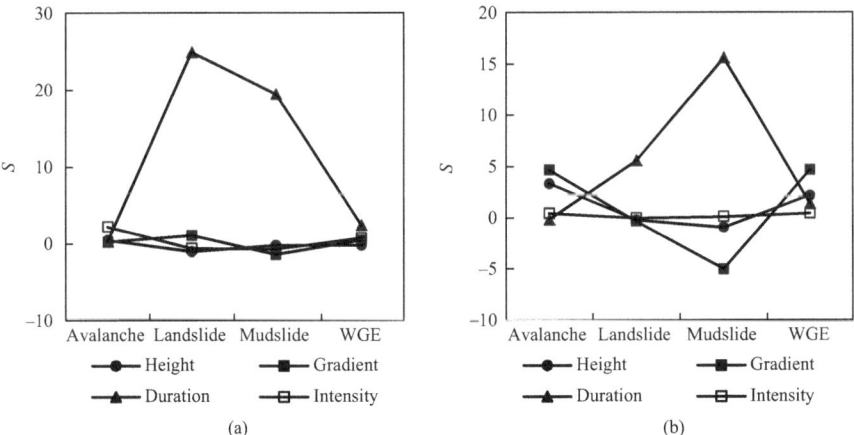

Fig. 10.11 Sensitivity analysis of the triggers. **a** Total amount; **b** Peak of the individual event

landslide and mudslide were 5.5 and 15.6, respectively. Meanwhile, the sensitivity parameters of the slope gradient on the avalanche and WGE were both 4.6. All of the parameters mentioned above were much larger than those of other triggers listed in the Fig. 10.11. It is worth mentioning that slope height was the second most influential element on the maximum gravity erosion, of which sensitivity parameters on the peak values of avalanche and WGE were 3.3 and 2.2, respectively. Nevertheless, the majority of the sensitivity parameters shown in Fig. 10.11 are less than or close to 1.0, which is much smaller than the maximum values above mentioned.

Accordingly, small changes in variables such as the duration and intensity of rainfall events could cause major changes in gravity erosion, demonstrating the significant sensitivity of these dynamics to climate change on the Loess Plateau. The result also suggests that rainfall duration-intensity and slope height-gradient are the prominent influential inputs on the steep loess slope. It is worth noting that both of these factors are potentially climate-driven: rainfall directly and landform through the effects of temperature and moisture on the soil's organic content (Sánchez-Canales et al. 2015). Collectively, a positive increment of those variables is associated with a positive increment of the failure amount and thus of the susceptibility level. If we compare the degree of vertical variability of the variable relief with the variable rainfall, we can observe that the strength of influence of the relief is lower than the rainfall.

For the sensitivity analysis, an increase-rate-analysis method was tested, which has never been previously employed in landslide susceptibility studies. As explained earlier, the aim of sensitivity analysis was to detect trends, interactions between variables, and non-linear behaviors. The approach allows a deeper understanding of the failure mechanism, especially when we want to recognize the effect of each causal factor and of the combination of factors on the susceptibility to gravity erosion. The method is easy to understand, facile to calculate, and its result is easy to read. It holds promise for research in other relevant fields of soil and water conservation.

10.4.6 Problems and Suggestions

This study will provide a tool for more reliable identification and analysis of site conditions associated with different types of failure. It is important, however, to note that although flume experiments are possibly the best approach to simulate natural slope failures, extrapolating these results to natural conditions requires caution because natural soil variability and initial stress conditions are difficult to reproduce in a flume study (Lourenço et al. 2006). Presently field tests on gravity erosion have been conducted by the authors in the Liudaogou Catchment of the Loess Plateau, and a comparison between the laboratory and field studies will be reported in the near future. The method of sensitivity analysis presented here is a novel way to evaluate the role of erosion triggers, especially in processing the results with limited data in the experimental study. However, the reliability of the analysis conclusions depends on the original data obtained in the laboratory or field. An in-depth qualitative review of the erosion rule will give great help to the test design and sensitivity analysis.

Vegetation also appeared to play a distinct role in controlling the locations of landslides at this site (Shakoor and Smithmyer 2005). In the vegetated areas, plant roots acted as a binder adding to soil strength, and thus reinforcing soils and improving their stability on slopes (Easson and Yarbrough 2002; Shields and Gray 1992; Waldron and Dakessian 1981). Thus experiments on the landform with different vegetation covers are expected in the future. Finally, the mechanical properties of soil, such as cohesion, friction angle, and density play an immediate role in the interpretation of mass failure behavior. Thus, a sensitivity analysis was also worth performing to evaluate variations in the factor of safety with respect to changes in the above mechanical parameters.

10.5 Conclusions

The experimental results obtained here provide insights into the pre-failure mechanisms and erosion processes of steep loess slopes. In an event of rainfall, various types of gravity erosion might emerge in the same period, and mass failures with the same mode and similar size often appeared adjacently in the same run of rainfall. Compound landslide, a group of mass failures might happen on a large slowly slipping block, was also observed in the tests.

Climate-driven factors and topography triggers had prominent influences on gravity erosion. Whether for the total amount or the peak amount in an experiment, the largest sensitivity parameter on both the landslide and mudslide was that of rainfall duration. In comparison, topography was relatively less influential. The sensitivity parameter of the slope gradient was the highest influence on the avalanche for the peak of an individual event.

References

Acharya G, Cochrane T, Davies T, et al. 2011. Quantifying and modeling post-failure sediment yields from laboratory-scale soil erosion and shallow landslide experiments with silty loess. Geomorphology, 129(1–2): 49–58.

Ali A, Huang J S, Lyamin A V, et al. 2014. Boundary effects of rainfall-induced landslides. Computers and Geotechnics, 61: 341–354.

Au S W C. 1998. Rain-induced slope instability in Hong Kong. Engineering Geology, 51(1): 1–36.

Benda L, Dunne T. 1997. Stochastic forcing of sediment supply to channel networks from landsliding and debris flow. Water Resources Research, 33(12): 2849–2863.

Cao Y Z. 1981. Mechanism and prediction of the gravity erosion on the loess area. Bulletin of Soil and Water Conservation, (4): 19–23 (in Chinese).

Easson G, Yarbrough L D. 2002. The effects of riparian vegetation on bank stability. Environmental and Engineering Geoscience, 8(4): 247–260.

Guzzetti F, Ardizzone F, Cardinali M, et al. 2009. Landslide volumes and landslide mobilization rates in Umbria, central Italy. Earth and Planetary Science Letters, 279(3–4): 222–229.

Haflidason H, Lien R, Sejrup H P, et al. 2005. The dating and morphometry of the Storegga Slide. Marine and Petroleum Geology, 22(1–2): 123–136.

Keefer D K, Larsen M C. 2007. Assessing landslide hazards. Science, 316(5828): 1136–1138.

Kirschbaum D B, Adler R, Hong Y, et al. 2010. A global landslide catalog for hazard applications: method, results, and limitations. Natural Hazards, 52(3): 561–575.

Li H Z, Yang Z S, Wang T R, et al. 2011. Recognition of landslide and case studies. Wuhan: Wuhan University of Technology Press: 19–43 (in Chinese).

Li T, Wang C, Li P. 2013. Loess deposit and loess landslides on the Chinese Loess Plateau// Shan W, Fathani T F. Progress of Geo-Disaster Mitigation Technology in Asia. Berlin Heidelberg: Springer: 235–261.

Liu C, Li W Y, Wu H B, et al. 2013. Susceptibility evaluation and mapping of China's landslides based on multi-source data. Natural Hazards, 69(3): 1477–1495.

Lourenço S D N, Sassa K, Fukuoka H. 2006. Failure process and hydrologic response of a two layer physical model: implications for rainfall-induced landslides. Geomorphology, 73(1): 115–130.

Melchiorre C, Castellanos Abella E A, van Westen C J, et al. 2011. Evaluation of prediction capability, robustness, and sensitivity in non-linear landslide susceptibility models. Guantánamo, Cuba. Computers and Geosciences, 37(4): 410–425.

Montgomery D R, Dietrich W E. 1994. A physically based model for the topographic control on shallow landsliding. Water Resources Research, 30(4): 1153–1171.

Parise M, Wasowski J. 1999. Landslide activity maps for landslide hazard evaluation: three case studies from Southern Italy. Natural Hazards, 20(2–3): 159–183.

Sánchez-Canales M, López-Benito A, Acuña V, et al. 2015. Sensitivity analysis of a sediment dynamics model applied in a Mediterranean river basin: global change and management implications. Science Total Environment, 502(1): 602–610.

Shakoor A, Smithmyer A J. 2005. An analysis of storm-induced landslides in colluvial soils overlying mudrock sequences, southeastern Ohio, USA. Engineering Geology, 78(3–4): 257–274.

Shields F D, Gray D H. 1992. Effects of woody vegetation on the structural integrity of sandy levees. Journal of the American Water Resources Association, 28(5): 917–931.

Tang K L. 2004. Soil and water conservation in China. Beijing: Science Press: 100–104 (in Chinese).

Thornes J B, Alcántara-Ayala I. 1998. Modelling mass failure in a Mediterranean mountain environment: climatic, geological, topographical and erosional controls. Geomorphology, 24(1): 87–100.

Waldron L J, Dakessian S. 1981. Soil reinforcement by roots: calculation of increased soil shear resistance from root properties. Soil Science, 132(6): 427–435.

Wang D F, Zhao X Y, Ma H L, et al. 1993. An investigation on the gravity erosion on the loess area. Soil and Water Conservation in China, (12): 26–28 (in Chinese).

Wu W M, Sidle R C. 1995. A distributed slope stability model for steep forested watersheds. Water Resources Research, 31(8): 2097–2110.

Xu X Z, Zhang H W, Xu S G, et al. 2009. Effects of dam construction sequences on soil conservation efficiency of a check-dam system. Journal of Beijing Forestry University, 31(1): 139–144 (in Chinese).

Xu C, Xu X W, Shyu J B H, et al. 2015a. Landslides triggered by the 20 April 2013 Lushan, China, Mw 6.6 earthquake from field investigations and preliminary analyses. Landslides, 12(2): 365–385.

Xu X Z, Liu Z Y, Wang W L, et al. 2015b. Which is more hazardous: avalanche, landslide, or mudslide? Natural Hazards, 76(3): 1939–1945.

Zhang M, Liu J. 2010. Controlling factors of loess landslides in western China. Environmental Earth Sciences, 59(8): 1671–1680.

Zhang D X, Wang G H, Luo C Y, et al. 2009. A rapid loess flowslide triggered by irrigation in China. Landslides, 6(1): 55–60.

Zhang R H, Liu X, Heathman G C, et al. 2013. Assessment of soil erosion sensitivity and analysis of sensitivity factors in the Tongbai-Dabie mountainous area of China. Catena, 101(2): 92–98.

Zhang C, Wang D G, Wang G L, et al. 2014. Regional differences in hydrological response to canopy interception schemes in a land surface model. Hydrological Processes, 28(4): 2499–2508.

Zhang A J, Zhang C, Chu J G, et al. 2015. Human-induced runoff change in Northeast China. Journal of Hydrologic Engineering, 20(5): 04014069.

Zhuang J Q, Peng J B. 2014. A coupled slope cutting—a prolonged rainfall-induced loess landslide: a 17 October 2011 case study. Bulletin of Engineering Geology and the Environment, 73(4): 997–1011.

Chapter 11
Detecting Fingerprints of Gravity Erosion Drivers: A Laboratory Experiment

Abstract The morphology of the failure scar has been a long-debated issue concerning the Loess Plateau of China, and the lack of normative data has hampered vital research in this area. In this chapter, a series of laboratory experiments were conducted to observe the failure geometries and volumes under rainfall simulations. The following six failure-scar types occurred: Tf, Cd, Cu, Ia, Ps, and Co. Tf, Cd, and Cu were the three major types of failure scars during the process of gravity erosion on a steep loess slope, and the scar Tf was the most significant. A relatively dangerous failure scar (Tf) might appear if the slope became steeper, or if the rainfall became more intensive. In addition, a long-duration storm could easily induce a large-volume failure with a Cu scar. The experimental results obtained here provide a morphogenic insight into the gravity-erosion control on a loess gully sidewall.

Keywords Loess Plateau · Gravity erosion · Failure scar · Rainfall · Topography

11.1 A Retrospective Study on the Failure Scar on the Slope

A failure scar is a crucial morphological characteristic for the reason that it can be regarded as a fingerprint to reflect the stability of a slope, the method of failure triggering, and the frequency and size of the gravity erosion. Previous papers (e.g., Zhang et al. 1997) reported that the failure scar could directly reflect the size of gravity erosion. Moreover, different morphologies of scars will correspond to distinct failure mechanisms (Millar and Quick 1997). Consequently, determining the morphologies of the scars is a key step for conserving soil and water and assessing the efficacy of gravity erosion treatment.

Gravity erosion, also called mass movement, mass wasting, slope movement, and slope failure (Blaschke et al. 2000), is the mass failure triggered by self-weight (Xu et al. 2015a) and is prone to occur on steep slopes (Xu et al. 2015b; Krohn et al. 2014; Basharat et al. 2014). Mass failures which are induced by rainfall frequently occur on the Loess Plateau because the area is characterized by complex topography of crisscrossing gullies having fragmentized terrain surface and extremely steep slope (Qiu et al. 2017). Numerous studies have suggested that the contribution of the mass failure to soil loss on the Loess Plateau is remarkable (Chen et al. 2007) and

© Science Press and Springer Nature Singapore Pte Ltd. 2020
X. Xu et al., *Experimental Erosion*,
https://doi.org/10.1007/978-981-15-3801-8_11

mass failure is the principal cause of long-term productivity losses (Luckman et al. 1999; Dregne 1995). In fact, mass failure processes essentially feature catastrophic removal or displacement by the gravity of the whole soil body from a slope. Hence, it frequently causes serious on-site and off-site damages. Local hazards, such as serious soil degradation, are led by the mass failure, because the mass failure often removes the entire soil profile in one event. Even if mass failure removes only part of the soil profile, that part almost always includes the organic matter and nutrient-rich A and upper B horizons (Blaschke et al. 2000). Mass failure also brings about off-site damages by sediment, such as transport route damage (Zieliński et al. 2016), building damage (Hungr et al. 2016), fluvial sediment deposition (West et al. 2014), reservoir sedimentation (Tsai et al. 2013), and channel silting (Sayed and González 2014). In addition, gravity erosion can cause fatalities. For example, a heavy rainfall-induced landslide in Xi'an on 17 September, 2011 caused 32 Fatalities (Zhuang and Peng 2014). The above-mentioned damages correlate with the pattern of the scar, because the failure scar (i.e., fingerprint) mirrors the size of the gravity erosion (Lucas et al. 2011).

The physics of fingerprinting has been embraced in studies of mass failure. For example, Convertino et al. (2013) argued that landslide size distribution is a fingerprint of the geomorphic effectiveness of rainfall as a function of climate change; Densmore and Hovius (2000) used the topographic fingerprints of bedrock landslides to distinguish triggers of earthquake and rainfall processes. Indeed, the application of the scars is the best way to explore the triggering mechanisms of mass failure since each hill slope generally produces a distinct geometry of scar, or fingerprint, during the process of mass failure. Previous studies have indicated that mass failures occurred in a natural slope with distinctly different morphologies of the scar. Skepton and Hutchinson (1969) showed that the ratio between depth and length of the failure scar (D_r/L_r) is generally between 0.15 and 0.33 when the mass slides along the upward concave scar. Meanwhile, the ratio of the scar is less than 0.1 when the mass slides along the translational face (Lu and Godt 2013). In addition, different morphologies of the scars are introduced in the stability analysis of the hillslope, such as planar, concave, convex, and terraced surfaces (Yin et al. 2009; Highland and Bobrowsky 2008; Leshchinsky et al. 1985). However, the lack of normative data tests has hampered the vital research in this area. Furthermore, varying failure surfaces were applied to the slope movement model, but all of them suffered from a significant lack of experimental validation and thus possibly led to an inaccurate prediction. Hence, it is necessary to correctly determine the scar morphology during the process of mass failure.

Several studies (e.g., Jeong et al. 2003; Lee et al. 1995) based on simplified scars reveal the mechanism of mass failure. However, applying the hypothetical scar, i.e., simplified scar shape, to study the triggering mechanism of gravity erosion might have inevitable limitations and flaws. Osman and Thorne (1988) assumed that steep slopes fail along an almost planar failure surface. The major limitation is that the failure plane is constrained to pass through the toe of the slope, which is unrealistic (Darby and Thorne 1996). Simon et al. (1991) also observed that the scar may, in fact, intersect the slope profile at other points. Moreover, the classical methods of slope stability analysis such as the Swedish slice method, Bishop method, and Janbu method, based on the conceptual arc-shaped scar to study the mass failure, also present some limitations. As addressed by Xu and Low (2006), analysis based solely upon circular slip surfaces may significantly overestimate the safety of the slope. In many cases, the failure scar may deviate significantly from a circle or a plane (Morgenstern and Price 1965). Therefore, the limitations simplifying the landform hampered attempts to apply the classical methods forecasting mass failures (Deng et al. 2011). In fact, the natural slip scars encompass a wide variety of surfaces that can satisfy any failure mechanism, such as the scar with differing shear strength properties or with complex pore-pressure distributions. Hence, to improve the accuracy and practicability of the model prediction, future projections should avoid too simplified conceptual scars and be based on a suitably complete mass-failure theory.

Sensitivity analysis plays an important role in exploring the triggering mechanisms of gravity erosion. The method helps to identify the most influential factors, including geology, geomorphology, and climate. The eigenvalues of erosive precipitation, such as rainfall amount, intensity, and duration, are the primary driving forces of gravity erosion on the Loess Plateau (Zhou et al. 2016; Zhou and Wang 1992), and the soil water content, infiltration depth, specific weight, cohesiveness, and friction angle change along with time during the rainfall infiltration period (Chang et al. 2013; Stark et al. 2005). Meanwhile, a slope failure is generally located along the slope of the stress concentration, while the geometry of the slope, such as slope shape, height, gradient, etc., is a critical factor affecting the stress distribution of the slope (Zhang and Fan 2015; Lu and Godt 2013). Global warming is expected to lead a more vigorous hydrological cycle, including more total rainfall and more frequent high-intensity rainfall events (Nearing et al. 2004). Thus, the above-mentioned factors are complex on the Loess Plateau, and are frequently affected by the rainfall-induced mass failures with different occurrences and sizes. Hence, identifying the most influential factor corresponding to the scar pattern may improve the accuracy and practicability of the prediction model of gravity erosion. Xu et al. (2015a) proposed an increase-rate-analysis method to recognize the effect of each causal factor and of the combination of factors on the susceptibility to gravity erosion. An advantage of the method is that the influences caused by the randomness of the gravity erosion can be readily overcome. Moreover, the method is easy to understand, simple to calculate, and its result is easy to read. Nevertheless, the average number of mass failures in an event of rainfall is not a suitable factor with which to analyze the sensitivity on the number of gravity erosion. Here, we improved the method using the total value (i.e., the total volume and number of mass failures in an event of rainfall) to evaluate the variations in the gravity erosion concerning the changes in various causal factors.

The objective of this study is to evaluate the natural factors affecting the distribution of scar morphology on the gully bank. To that end, we conducted a series of steep slope collapse experiments, and we used a topography meter designed by us to observe the patterns of failure scars in the process of mass failure. Moreover, the increase-rate-analysis method was used to analyze the sensitivity of gravity erosion corresponding to different scar morphologies.

11.2   Methods and Materials

The gully bank collapse experiments were conducted under closely controlled conditions in the Joint Laboratory for Soil Erosion of Dalian University of Technology and Tsinghua University located in Beijing, China. The landscape simulator consisted of a rainfall simulator and a conceptual landform covering an area of 3.0 m by 3.0 m (Fig. 10.2). The conceptual landforms with a gentle upper slope of 3° were made with loess by hand patting. More details of the gully bank collapse experiments (slope height-gradient, rainfall intensity-duration) are provided in Table 11.1. Five runs of rainfall were applied, in turns, on a landform, and the rainfall interval was approximately 12 h. Furthermore, the physical properties of the model soil—the 50% diameter of soil particles (d_{50}) and the specific gravity (γ_s), were 0.05 mm and 2.56, respectively.

The form of a failure scar (i.e., the geometry of slip surfaces and the shape of the vertical profiles on the slope surface) is created as a result of the soil block separating from a sloped face. Herein, we used the type of intersection line to define the morphology of failure scar in the experiments. The line which ran through the intersection of the failure scar and the longitudinal section which passed through the center of the scar was obtained.

During, and 20 min after, the rainfall, slope failure occurrence time, slip mode, type of failure scar, location, and slope failure retrogression behavior were recorded by direct observation and via the topography meter designed by us (Guo et al. 2016). To ensure the accuracy of the scar pattern obtained from direct observation during the experiments, the videos in the topography meter were used to check the pattern.

For each mass failure, we calculated the volume and classified the scar form. Then, we obtained the total amount of soil loss (g_n) and the sum amount of mass failures (g_v) corresponding to all types of scars. All failure masses with volume greater than 500 cm^3 were considered in the study. To evaluate the independent effects of slope gradient and height, and rainfall intensity and duration on the mass volume of each scar pattern, for this study the experiments were divided into the following seven experimental groups. Each experimental group had the same slope height or gradient, and rainfall intensity or duration:

(1) G_a (experiments L1, 2, 5 and 6) versus G_b (experiments L3, 4, 7 and 8). Slope height of the initial lower slope in the former experimental group was 1.0 m, while the latter was 1.5 m.

(2) G_c (experiments L1, 3, 5 and 7) versus G_d (experiments L2, 4, 6 and 8). Slope gradient of the initial lower slope in the former experimental group was 70°, while the latter was 80°.

(3) G_e (experiments L5–6) versus G_f (experiments L9–10). Rainfall intensity in the former experimental group was 0.8 mm/min, while the latter was 2.0 mm/min.

(4) G_f (experiments L9–10) versus G_h (experiments L1–2). Rainfall duration in the former experimental group was 30 min, while the latter was 60 min.

Then we used the increase-rate-analysis method to assess variations in the gravity erosion in regard to changes in other causal factors such as rainfall intensity and

Table 11.1 Volume and number of mass failures with different scar patterns for an initial landform after five rainfalls

Test number	Number of mass failures						Total volume of mass failures (10^3 cm³/m)					
	Tf	Co	Uc	Dc	Ps	Ia	Tf	Co	Uc	Dc	Ps	Ia
L1(S30-1-70d)	18	0	5	16	2	6	31.9	0.0	3.3	18.1	7.3	17.2
L2(S30-1-80d)	11	0	6	8	1	0	213.8	0.0	50.8	64.9	1.8	0.0
L3(S30-1.5-70d)	30	0	25	11	1	1	107.7	0.0	82.5	7.2	2.4	2.3
L4(S30-1.5-80d)	11	1	11	6	0	0	191.1	1.2	79.7	19.1	0.0	0.0
L5(G60-1-70d)	10	0	2	8	0	0	123.2	0.0	15.3	11.2	0.0	0.0
L6(G60-1-80d)	12	0	8	5	1	1	41.3	0.0	105.6	24.7	20.1	7.6
L7(G60-1.5-70d)	20	1	2	12	1	1	112.3	7.9	3.8	82.1	0.7	0.5
L8(G60-1.5-80d)	15	0	5	1	0	0	171.4	0.0	13.5	13.4	0.0	0.0
L9(G30-1-70d)	4	1	1	3	1	0	30.1	1.7	0.1	9.8	20.4	0.0
L10(G30-1-80d)	3	0	4	6	0	0	7.5	0.0	8.9	36.5	0.0	0.0
Total	134	3	69	76	7	9	1030.3	10.7	363.4	287.0	52.7	27.6
Percentage (%)	45	1	23	26	2	3	58	1	21	16	3	2

duration, and slope gradient and height. The sensitivity analysis focuses on g_n and g_v. The increase ratios of R_g (%) are obtained as follows:

$$R_g = (g_2 - g_1)/g_1 \tag{11.1}$$

where g_1 is the total volume of all failures, g_n, or the individual volume each event, g_v, for an experiment before the triggering element was changed, cm^3; g_2 is that after the triggering element was changed, cm^3. An increased ratio of the above output in percent is calculated with the growth rate of a parameter, while other parameters are fixed in an experiment:

$$S = R_g/R_t \tag{11.2}$$

$$R_t = (t_2 - t_1)/t_1 \tag{11.3}$$

where S is the sensitivity parameter to analyze the sensitivity of the triggering elements on the failure number and volume of each scar morphology; t is one of the triggering elements, e.g., slope height, slope gradient, rainfall duration, and rainfall intensity; R_t is the increased ratio of the triggering element, %; t_1 is the value before being changed in an experimental group; t_2 is that after being changed. This approach allows for a qualitative investigation of the effect of conditioning factors.

11.3 Results and Discussion

11.3.1 Characteristics of Failure Scars

The classification of failure-scar morphology is valuable because it can reflect the stability of slope scar, the triggering method of the mass failure, and the frequency and size of the gravity erosion. We observed the scar morphologies of all gravity erosions in ten sets of gully bank collapse experiments (Table 11.1). Then, we classified the scars into the following six types based on the shapes of the intersection lines:

(1) Translational face (Tf). The scar morphology, which is an inclined plane with an intersection line consisting of a straight line, is termed the translational face (Fig. 11.1a). This implies that the soil block fails along an almost planar failure surface.

(2) Polygonal side (Ps). The scar morphology is called a polygonal-side when the intersection line is a broken line. This kind of scar has at least two intersecting planes, one of which is the low-angle incline plane, and another which is a relatively steep surface. As shown in Fig. 11.1b, the Ps scar closely resembles the shape of a stair-step.

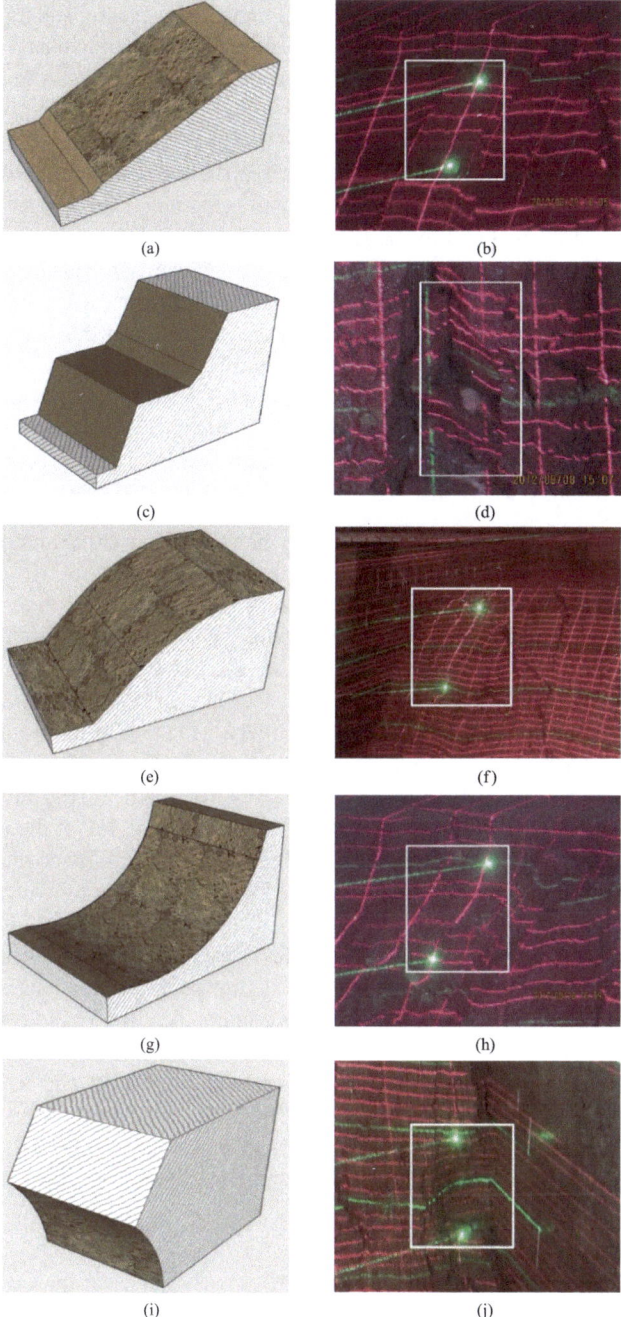

Fig. 11.1 Schematic representations and sample photos of the scar morphologies. The figure contains the schematic representations on the left and the corresponding observed exemplar on the right. The intersection lines are shown in the left graphs. **a**, **b** Translational face; **c**, **d** Polygonal-side; **e**, **f** Convex; **g**, **h** Upward concave; **i**, **j** Downward concave

(3) Convex (Co). If the intersection line is approximating an arc, and its corresponding center is in the inner slope, the scar morphology is known as a convex scar. The Co scar, characterized by a gentle upper slope and steep lower slope, is given in Fig. 11.1c.

(4) Upward concave (Uc). The Uc scar is defined as the intersection line that is close to an upward concave arc. This means that the scar is an upward-arc surface with a reverse foreside, gentle middle, and steep tail. The Uc scar is similar to the crescent-shaped bed form, as shown in Fig. 11.1d.

(5) Downward concave (Dc). The Dc scar is a downward-arc surface. Morphologically speaking, a Dc scar is different from an Uc scar for the reason that the center of the downward-arc is placed in the downside of the scar. Generally, a Dc scar has lower stability compared with a Uc scar. Figure 11.1j shows a Dc scar, which looks like a spoon-shaped cavity or parabolic cavity on the slope.

(6) Irregular appearance (Ia). A few irregular appearances also showed up in the experiments; they were composed of at least two kinds of surfaces mentioned in the previous paragraphs.

As shown in Table 11.1, ten sets of gully bank collapse experiments were conducted at the slope heights of 1 and 1.5 m, and slope gradients of $70°$ or $80°$ under the rainfall intensities of 0.8 and 2 mm/min, respectively. Experimental results show that the number of mass failures of the scars Tf, Dc, Uc, Ia, Ps, and Co were 134, 76, 69, 9, 7, and 3, respectively. That is, the number of mass failures in the experiments corresponding to the scars Tf, Dc, Uc, Ia, Ps, and Co amounted to 45, 26, 23, 3, 2 and 1% of the total, respectively. Meanwhile, the total volume of failures for six distinct morphological types, namely Tf, Uc, Dc, Ps, Ia, and Co were 1030.2×10^3, 363.4×10^3, 287.0×10^3, 52.7×10^3, 27.6×10^3, and 10.8×10^3 cm^3/m, and their volumes accounted for 58, 20, 16, 3, 2 and 1% of the total, respectively. This fact implies that Tf, Uc, and Dc were three types of major scars occurring in the processes of gravity erosion on the steep loess slope, among which the Tf was the most crucial.

The large-scale and frequent mass failures may harm both local buildings and lives. In the experiments shown in Fig. 11.2, the peak volume of the individual mass failures with the Tf, Uc, and Dc scars accounted for 78, 17, and 5% of the total amount of whole peak events. Meanwhile, the frequencies of mass failure occurrences corresponding to Tf, Uc, and Dc amounted respectively to 60, 30, and 10% of the total number in all of the maximum-mass failures. Consequently, on the steep slope made by hand with loess and under intense rainstorms, hazards caused by the translational-shaped failures may be much more serious and hazardous than those caused by the other shaped failures.

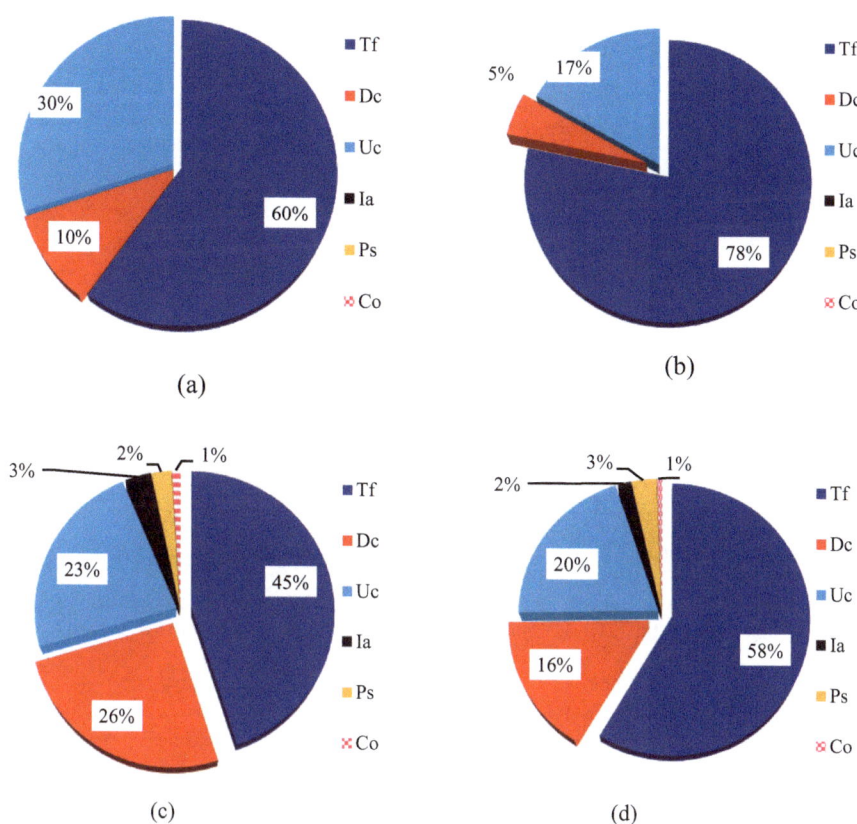

Fig. 11.2 Distribution of the scar morphologies in the maximum of the individual failure masses in ten sets of experiments. **a** The frequencies of mass failures with the scars translational face (Tf), upward concave (Uc), downward concave (Dc), convex (Co), polygonal-side (Ps), and irregular appearance (Ia) in all maximum mass failures. **b** The proportion of the largest volumes of the individual mass failures with the scars Tf, Uc, Dc, Co, Ps, and Ia in all peak events. **c** The frequencies of mass failures with the scars Tf, Uc, Dc, Co, Ps, and Ia in all mass failures. **d** The proportion of volume of mass failures with the scars Tf, Uc, Dc, Co, Ps, and Ia in all events. Although the peak volume of the individual mass failures has not been observed in Co, Ps, and Ia, in fact, those types of mass failures also occurred in the experiments

11.3.2 Impact Factors of Failure Scars

To assess the effects of topography and rainfall on scar pattern, the frequency and size of gravity erosions of each scar pattern in every group experiment were compared, as shown in Figs. 11.3 and 11.4. The histogram in Fig. 11.3a shows the variations of number and volume for the mass failures corresponding to six types of scars in an event of rainfall as the slope height was increased. While other factors were fixed and the slope height was increased from 1.0 to 1.5 m, the total number and amount of failures for the scar Tf were increased by 49 and 42%, respectively. Meanwhile, the

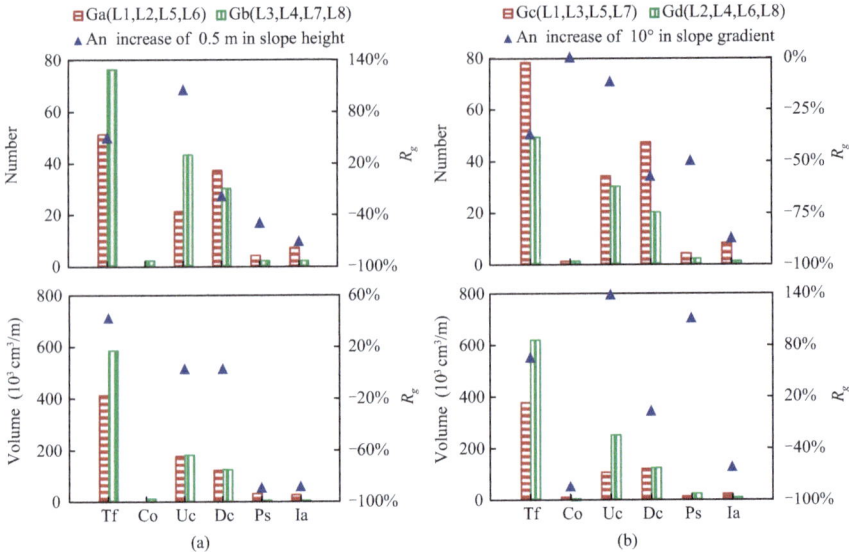

Fig. 11.3 Increment of the gravity erosion with the increase of slope height and gradient. **a** Slope height was increased from 1.0 to 1.5 min; **b** Slope gradient was increased from 70° to 80°

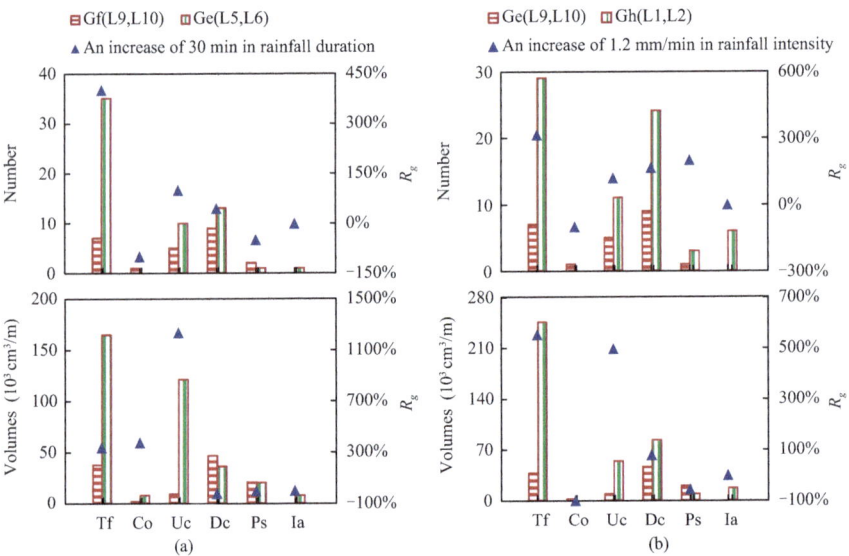

Fig. 11.4 Increment of the gravity erosion with increase of rainfall duration and intensity. **a** Rainfall duration was increased from 30 to 60 min; **b** Rainfall intensities were increased from 0.8 to 2 mm/min

total number of Uc scars was enhanced by 105%, and the mass failure with the scar Dc was reduced by 19%, but their amounts did not change significantly. In addition, the total number and amount of mass failures corresponding to the scars Ps and Ia were decreased by more than 50%. However, for a slope with the height of 1.0 m or 1.5 m, the probability of the Co scar occurrence was low in two-group comparisons.

Slope gradient also has an important influence on the distribution of the scar morphologies. As shown in Fig. 11.3b, when other parameters were fixed but the slope gradient was increased by 14% (i.e., the slope gradient was increased from 70° to 80°), the total numbers of mass failures for the six different types of scars mentioned above were decreased, whereas their volumes were increased, with the exception of Co and Ia scars that were reduced. This phenomenon can be attributed to the following two reasons. First, the increased slope gradient caused the bad infiltration on the slope face, in turn resulting in a lower frequency of mass failures for the scars. Table 11.1 provides strong evidence that the occurrence frequency of mass failure was decreased as the slope gradient increased. For the gully banks with a slope of 70° (i.e., the experiments L1, 3, 5, 7 and 9), the total number of mass failures was 182, while for those with a slope of 80° (i.e., the experiments L2, 4, 6, 8 and 10), the total number of mass failures was 116. Second, the slope stability decreased with the rise of slope gradient. As shown in Table 11.2, in the experiments group with the slope of 70° (i.e., the experiments L1, 3, 5, 7 and 9), the total amount of all peak individual mass failures was 606.22×10^3 cm^3. Nevertheless, the total volume of all peak events for the experiments L2, 4, 6, 8 and 10 (with the slope of 80°) was 914.95×10^3 cm^3. That is to say, the total peak volume of the individual mass failures was augmented by 51% as the initial slope increased from 70° to 80°. In addition, during rainfall as the soil moisture content rose, ultimately leading to

Table 11.2 The maximum volume of the individual failure masses with different scar patterns in the experiments

Test number	Maximum volume of individual failure events (10^3 cm^3)					
	Tf	Uc	Dc	Co	Ps	Ia
L1(S30-1-70d)	–	41.13	–	–	–	–
L2(S30-1-80d)	168.41	–	–	–	–	–
L3(S30-1.5-70d)	–	97.16	–	–	–	–
L4(S30-1.5-80d)	369.91	–	–	–	–	–
L5(G60-1-70d)	177.6	–	–	–	–	–
L6(G60-1-80d)	–	117.7	–	–	–	–
L7(G60-1.5-70d)	224.06	–	–	–	–	–
L8(G60-1.5-80d)	183.71	–	–	–	–	–
L9(G30-1-70d)	66.27	–	–	–	–	–
L10(G30-1-80d)	–	–	75.22	–	–	–
Total	1189.96	255.99	75.22	–	–	–
Percentage (%)	78	17	5	–	–	–

large-scale mass failures, the peak events occurred during the third or fourth rainfall events for five runs of rainfalls in each experiment.

The effect of rainfall duration was also investigated by varying the rainfall duration from 30 to 60 min, while the rainfall intensity was kept constant at 0.8 mm/min. Figure 11.4a portrays the changes of the aforementioned failure scars as the rainfall duration increased. The number of gravity erosions corresponding to the Tf, Uc and Dc were increased, among which Tf was significantly augmented by 400%, while gravity erosions with the scars Co and Ps were reduced by more than 50%. Meanwhile, the volumes of gravity erosions corresponding to the scars Tf, Uc and Co were increased by 338, 1236 and 376%, respectively, but no substantial change occurred in the failures with the scars Dc and Ps. In other words, the change of volumes for different scars was tremendous, and the volume of the Uc-shaped failure underwent a significant increase.

Rainfall intensity also has a direct impact on gravity erosion. Inspection of Fig. 11.4b revealed that there had been a very substantial increase in the total number or amount of gravity erosions corresponding to three major types of scars (i.e., Tf, Uc, and Dc) when the rainfall duration was kept constant at 30 min but the rainfall intensity was increased from 0.8 to 2.0 mm/min. In particular, the volume of gravity erosion for the scar Tf was augmented by 553%. This increase indicates that rainfall intensity significantly influenced the size and frequency of gravity erosion of the major scars. In contrast, the changes in the gravity erosion corresponding to the scars Co, Ps, and Ia were not statistically significant.

11.3.3 Sensitivity Analysis

Here, a sensitivity analysis was implemented to evaluate the influences of topography and rainfall on the size and frequency of the mass failures corresponding to the six forms of scars. As shown in Fig. 11.5a, for the number of mass failures, the

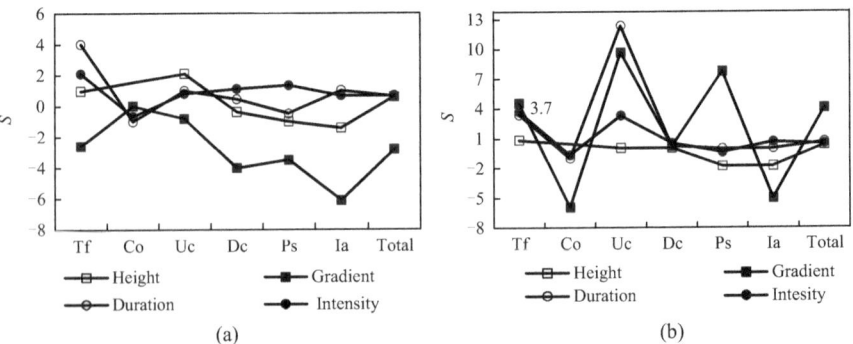

Fig. 11.5 Sensitivity analysis of the impact factors. **a** Total number; **b** Total amount

sensitivity parameters such as slope height, slope gradient, rainfall duration, and rainfall intensity for the scars Tf, Dc, Uc, Ia, Ps, and Co ranged from −6.1 to 3.7. In particular, the rainfall duration had significant effects on the number of failures with the scar Tf ($S_{NTf} = 4.0$). The frequency of mass failures with the scars Dc, Ps, and Ia were greatly influenced by the change of slope gradient, and their sensitivity parameters were −4.0, −3.5, and −6.1, respectively. In other words, the number of mass failures with the scar Uc ($S_{NUc} = 2.1$) was highly susceptible to slope height. Meanwhile, for the volume of gravity erosion in the experiments, the results of sensitivity analysis (Fig. 11.5b) demonstrated that the sensitivity parameters of slope height, slope gradient, rainfall duration, and rainfall intensity on the scars Tf, Dc, Uc, Ia, Ps and Co varied dramatically, with values ranging from −5.9 to 12.4. It is worth mentioning that rainfall duration was the most influential element on the size of mass failures, for the sensitivity parameter of the rainfall duration on the total volume of the scar Uc was 12.4. The slope gradient was the second most prominent trigger factor on the size of the Uc-shaped failure, and the value of the sensitivity parameter was 9.7. Additionally, the sensitivity parameters of the slope gradient on the total volume of failures with the scars Tf, Co, Ps and Ia were comparable, at 4.5, 5.9, 7.8 and 5.0, respectively. In addition, the sensitivity parameters of slope gradient on the total volume and number of gravity erosions were 2.8 and 4.1, respectively.

11.3.4 Formation Mechanism of Scar Morphologies

Loess is a special type of geological material which has soil structure homogeneity and peculiar mechanical properties, leading to the failure scar characterized by obvious geometrical morphology (Gan et al. 1999). In this study, the scar pattern also shows a distinct geometric appearance, which could be divided into six types (i.e., Tf, Uc, Dc, Ps, Co, and Ia). Generally, the mass failure occurs along an inclined plane when the shear strength is primarily provided by the inter-particle frictional resistance (Wang and Li 2009). Although the loess is sticky in the experiment ($d_{50} = 0.05$ mm), the cohesion rapidly decreases to zero while the internal friction angle normally reaches a stable state if the moisture content exceeds the plastic limit (Derbyshire et al. 1994). Thus, the soil mass falls along the scar Tf. As we know, the mass failure strongly depends on the shear strength of the slope material, which is not only related to the internal friction angle but also the cohesion. Change of the apparent cohesion due to variations in the soil water content results in a re-distribution of shear strength during the process of rainwater infiltration. Owing to the phenomenon that the strength diminishes as shear stress increases, the mass failure process is characterized by a non-uniform distribution of the shear strength along the potential slip surface (Chen et al. 2016). Consequently, rainfall-induced failures also bring about different forms of scars. For instance, the failure will also produce an arc-shaped scar, including Uc, Dc, and Co, when the cohesion plays an important role in the shear strength of the soil (Wang and Li 2009). However, the proportion of cohesion in the shear strength is dynamic in the study. As a result, other scar morphologies occurred, including Tf, Ps, and Ia. There is significant evidence that the mass failure depends critically on the shear strength, tensile strength, and effective stress of the slope material (Goulding 2006). In addition, although the initial slope gradients were 70° or 80° in the experiments, the instantaneous slope gradients of the individual mass failures were different, as the failures frequently occurred. Hence, the dynamic slope gradients might affect the shear strength of the slope surface (Meng 1996). However, rainwater infiltration triggers the variations in the soil water content, leading to increased pore water pressure, and in turn causing re-distribution of the macroscopic stress and strength of the potential scar, together with the dynamic changes of instantaneous slope and the cohesion, and ultimately resulting in different scar morphologies.

11.3.5 Effects of Parameters on Scar Forms

Failure scars could reflect the method of failure triggering, and the frequency and size of the gravity erosion (Zhang et al. 1997; Millar and Quick 1997). In the study, Tf, Uc, and Dc were the three major types of failure surfaces occurring in the processes of gravity erosion on the steep loess slope, among which the Tf was the most crucial. The result is close to other similar studies on the Loess Plateau. Shi et al. (2016) also found that the arc-shaped scar and the translational scar were two major types of scars in the northern bank of the Weihe River on the Loess Plateau. In fact, most of the well-preserved arc-shaped scars on the natural slopes are Uc scars, because the stability of the Dc scar is weak, which tends to induce subsequent mass failures. Meanwhile, profound and obvious discrepancies were seen among the mass failures for different scars in the experiments. Indeed, changing the scar geometry while keeping the upper surface of the released mass constant induces a change in the released volume (Lucas et al. 2011). Our study reveals that the total volume of the failure masses with the scars Tf, Uc, and Dc accounted for a large percentage of the total volume for the mass failures, of which the Tf was the most decisive.

Different triggering methods can explain why the failures volume of the Tf was larger than those of the Uc and Dc. In this study, there were three ways to trigger mass failures, including the method of crack propagation (Fig. 11.6), the deformation of partially saturated soil (Fig. 11.7), and the combined effects of the crack propagation and deformation of partially saturated soil (Fig. 11.8). In comparison, the third triggering approach (i.e., the method driven by the combined effects of crack propagation and saturated soil), which was prone to form a Tf scar, might result in the maximum volume in the experiments. A failure scar is not only a fingerprint of the slide trace or a vertical profile formed by the soil block separating from a sloped face (Waldmann et al. 2011). It is also a measure of geomorphic effectiveness of rainfall as a function of climate change (Convertino et al. 2013). In turn, a mass failure is affected and constrained by rainfall parameters, particularly the rainfall intensity and duration (Guzzetti et al. 2008).

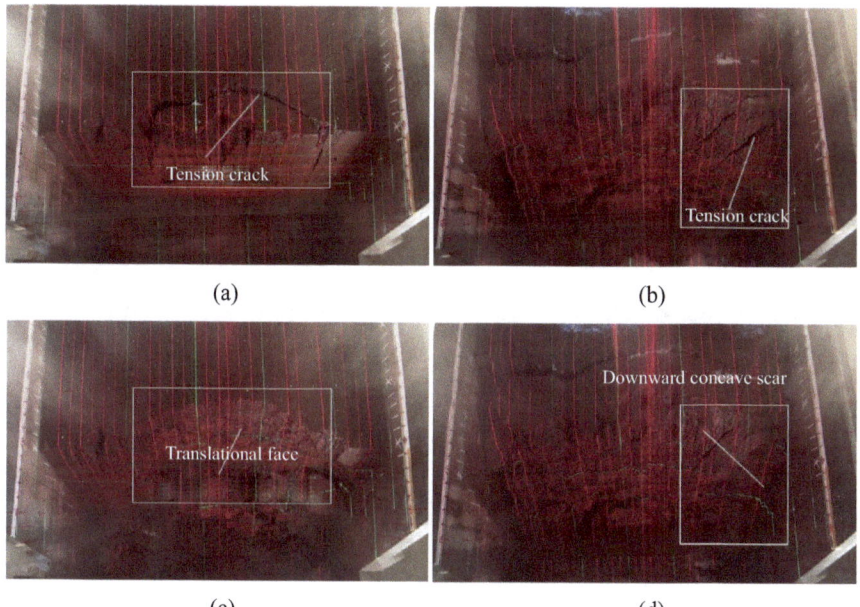

(a)   (b)

(c)   (d)

Fig. 11.6 Triggering method of crack propagation. **a, b** Tension cracks were created and expanded, which indicated a landslide was coming; **c** Translational face; **d** Downward concave scar. Here is a failure scar caused by propagation of cracks

Our results show that the rainfall duration significantly influenced the size and frequency of the mass failure. Furthermore, the failure with the scar Tf or Uc was increased dramatically (Fig. 11.5a). In addition, the results of this study reveal that, with an increase in rainfall intensity, both the number and volume of the mass failures of the three main types of scars (i.e., Tf, Uc, and Dc) were augmented (Fig. 11.4b). Previous research showed that with an increase in the landslide volume, the most important rainfall variable might be the rainfall duration (Dai and Lee 2001). In addition, our results show that the rainfall intensity significantly influenced the size and frequency of gravity erosions with the different scars (Fig. 11.4b). As the rainfall intensity rose, the frequency of mass failures was increased, likely because the intensive rainfall led to a high initial and steady infiltration rates, as well as a rapid increment in the water infiltration depth and accumulated infiltration (Li and Shao 2006).

Topography parameters, such as the slope height and gradient, also have important influences on the distribution of scar morphologies. In this study, as the slope height grew, the events and amounts of the three major types of failures (i.e., the failures with the scars Tf, Uc, and Dc) were augmented. In particular, the number of Uc-shaped failures and the number and volume of Tf-shaped mass failures were significantly increased (Fig. 11.3). Increasing the slope height, which encourages the shear stress concentrations arising at the toe and tension stress at the top, will result

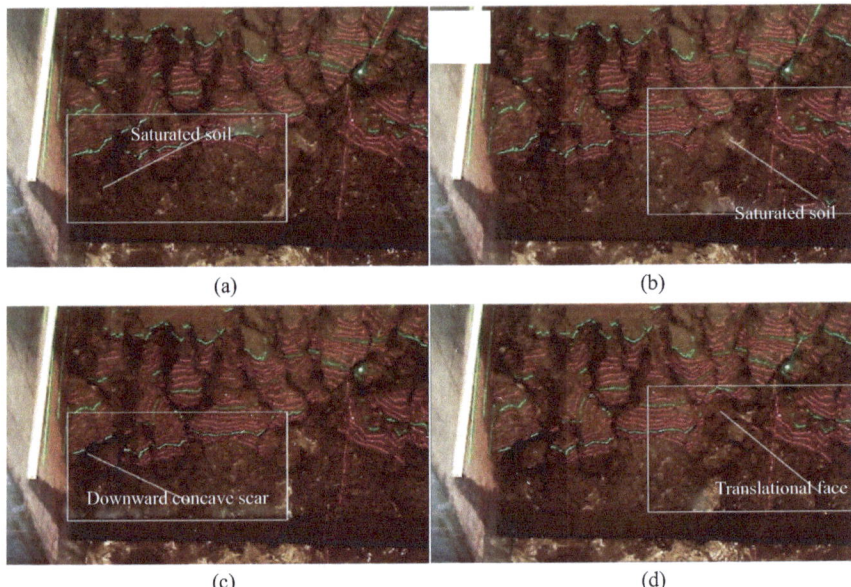

Fig. 11.7 Triggering method of partially saturated soil deformation. **a, b** Partially saturated soil becomes more saturated, leading to the deformation of saturated soil at the bottom, and followed by slope failure; **c** Downward concave scar; **d** Translational face. Here is a failure scar caused by deformation of the partially saturated soil

Fig. 11.8 Triggering mode driven by the combined effects of tension crack and saturated soil. A Tf scar was formed here

in a significantly increased probability of mass failure. In addition, slope height had a striking effect on mass failure size and frequency (Qiu et al. 2017). As the slope gradient increased, the total amount of mass failure with the Tf, Dc, and Uc scar were increased (Fig. 11.3b). The results agree with the simulation of Katz et al. (2014) which revealed that the large-scale mass failures are associated with higher slope gradient.

In comparison, the frequency and size of the mass failure with different scar patterns were more susceptible to the change of slope gradient than they were to height (Fig. 11.5). Zhu and Hong (2011) also discovered that the slope gradient was more sensitive to the slope stability than to slope height. Slope gradient determines the state and distribution of stresses in the slope mass, and controls the stability and mode of instability of the slope (Dai et al. 2017; Zhang and Liu 2010). In addition, slopes with lower gradients required more cumulative precipitation before a mass failure occurred (Wu et al. 2017). For the rainfall characteristics, rainfall intensity and duration were equally important in influencing the frequency of mass failure with the scar Uc, while the rainfall duration significantly affected its size (Fig. 11.5a, b). This could explain why shallow landslides are typically induced by intense rainfall, while deep-seated landslides require rainfall with a longer duration (Zhang et al. 2016). It is the increased rainfall duration that increases the water content and pore-water pressure and thus reduces the soil shear strength, eventually resulting in mass failure (Xu et al. 2013). Meanwhile, the rainfall intensity affects the rainfall kinetic energy, which shortens the pathway for water to approach deeper soil strata, thus increasing the possibility of mass failure (Lin and Chen 2012).

11.3.6 Hazard Risks Suggested from the Scars

The damage caused by a mass failure largely depends on its velocity and volume, but the velocity is extremely difficult to evaluate systematically. Hence, some researchers assessed the hazard of mass failure using the landslide type, volume, and scar (e.g., Jaiswal et al. 2011). In this study, the mass volume and scar morphology were applied to assess the damage caused by gravity erosion. The result of the study is that the damage caused by the translational-shaped failure may be much more serious than that caused by the other shaped failures. Particularly, the large-scale rapid mass failures corresponding to the Tf scar occurred on the steep slope, and it was often accompanied by a crack at the top of the slope (Fig. 11.8). Due to the crack occurrence on the top of the slope, the mass failure with the Tf scar will widen the channel. In other words, once gullies develop, they increase the connectivity in the landscape, hence raising the risk of flooding and reservoir sedimentation in a catchment (Ionita et al. 2015; Poesen et al. 2003). Derbyshire (2001) also pointed out that the mass failure with the scar Tf has the characteristics of rapid disintegration and high sliding velocities, and this kind of failure is abundant in the loess region of North China. As we know, rapid landslides pose lethal threats, whereas slow movements damage properties but seldom cause fatalities (Iverson et al. 2000). Thus, the mass failure with the scar Tf, which moves quite rapidly downhill, is the most violent gravity erosion and might cause great harm to buildings and lives.

Although the slope gradients of the initial landform were 70° or 80° in the experiments, multiple mass failures occurred in one region, resulting in a decrease of the slope gradient. Therefore, the arc-shaped scars, including Uc, Dc, and Co, also occurred in the experiments. In the experiments L1, L3, and L6, the scar morphology with the maximum volume of the individual mass failures was the Uc scar, although in the experiment L10 it was the Dc scar (Table 11.2). Generally, the above-mentioned individual mass failures occurred during the third or fourth rainfall events for each model. This implies that multiple mass failures after two or three rainfalls may decrease the initial slope gradient and lead to the occurrence of mass failure with the Uc and Dc scars. Indeed, the mass failures with Dc and Uc always occur on the relatively gentle slope rather than a nearly vertical slope. Lohnes (1991) also found that the mass failure with the scar Tf always occurred on steep gully banks, and that with the arc-shaped scar (Uc and Dc) they generally happened at the low slope angles where the surfaces of slab or planar failure unlikely exist. Moreover, a Dc scar with low stability tends to cause subsequent mass failures. For example, the concave-shaped failure generally formed a spoon-shaped cavity on the slope, as shown in Fig. 11.1j, and then the soil block on top of the cavity surface may fall due to its own gravity.

In addition to the damages mentioned above, the mass failure also caused the channel to widen because mass failure was a frequent and successive process of geomorphic evolution. If the mass failures on the gully bank were not controlled, the process of gully expansion would reach its maximum size, forcing farmers to retreat and reduce the cultivated area around the gullies (Yitbarek et al. 2012). From this perspective, gravity erosion is an important geomorphic natural hazard that affects livelihoods in a catchment, especially on the Loess Plateau of China.

11.4 Conclusions

The experimental results obtained here provide a morphogenic insight into a mechanism analysis of the gravity erosion on the gully banks. In an event of rainfall, various scar morphologies of mass failure might emerge in the same period. The scars Tf, Uc, and Lc were the three major types of scars occurring in the processes of gravity erosion on the steep loess slope, among which the Tf was the most crucial.

Climate-driven factors and topography triggers prominently influenced the distribution of the scar morphologies. Whether for the total amount or total number in the experiments, the largest sensitivity parameter of the influential factors was rainfall duration. The distribution of the scar pattern was highly susceptible to the slope gradient. In comparison, slope height was relatively less influential.

References

Basharat M, Rohn J, Baig M S, et al. 2014. Spatial distribution analysis of mass movements triggered by the 2005 Kashmir earthquake in the Northeast Himalayas of Pakistan. Geomorphology, 206(1): 203–214.

Blaschke P M, Trustrum N A, Hicks D L. 2000. Impacts of mass movement erosion on land productivity: a review. Progress in Physical Geography, 24(1): 21–52.

Chang T H, Lin C L, Huang F K. 2013. Slope cracks in the safety assessment of the rainfall period. Applied Mechanics and Materials, 405–408(2): 2364-2369.

Chen L D, Wei W, Fu B J, et al. 2007. Soil and water conservation on the Loess Plateau in China: review and perspective. Progress in Physical Geography, 31(4): 389–403.

Chen X P, Zhu H H, Huang J W, et al. 2016. Stability analysis of an ancient landslide considering shear strength reduction behavior of slip zone soil. Landslides, 13(1): 173–181.

Convertino M, Troccoli A, Catani F. 2013. Detecting fingerprints of landslide drivers: a MaxEnt model. Journal of Geophysical Research: Earth Surface, 118(3): 1367–1386.

Dai F C, Lee C F. 2001. Frequency-volume relation and prediction of rainfall-induced landslides. Engineering Geology, 59(3–4): 253-266.

Dai F, Li B, Xu N W, et al. 2017. Microseismic monitoring of the left bank slope at the Baihetan hydropower station, China. Rock Mechanics and Rock Engineering, 50(1): 225–232.

Darby S E, Thorne C R. 1996. Development and testing of riverbank-stability analysis. Journal of Hydraulic Engineering, 122(8): 443–454.

Deng D P, Li L, Zhao L H. 2011. A new method for analysis of slope stability under steady seepage. Journal of Engineering Geology, 19(1): 29–36 (in Chinese).

Densmore A L, Hovius N. 2000. Topographic fingerprints of bedrock landslides. Geology, 28(4): 371–374.

Derbyshire E. 2001. Geological hazards in loess terrain, with particular reference to the loess regions of China. Earth-Science Reviews, 54(1): 231–260.

Derbyshire E, Dijkstra T A, Smalley I J, et al. 1994. Failure mechanisms in loess and the effects of moisture content changes on remoulded strength. Quaternary international, 24: 5–15.

Dregne H E. 1995. Erosion and soil productivity in Australia and New Zealand. Land Degradation and Development, 6(2): 71–78.

Gan W J, Huang Y H, Zhang P Z. 1999. Method for determining the most risk slip surface of loess slope by genetic algorithm and its application to stability analysis. Journal of Engineering Geology, 7(2): 168–174 (in Chinese).

Goulding R B. 2006. Tensile strength, shear strength and effective stress for unsaturated sand. Columbia: University of Missouri.

Guo W Z, Xu X Z, Wang W L, et al. 2016. A measurement system applicable for landslide experiments in the field. Review of Scientific Instruments, 87(4): 044501.

Guzzetti F, Ardizzone F, Cardinali M, et al. 2008. Distribution of landslides in the Upper Tiber River basin, central Italy. Geomorphology, 96(1–2): 105-122.

Highland L M, Bobrowsky P. 2008. The landslide handbook-A guide to understanding landslides: Reston, Virginia, USGS Science for a changing world-Circular 1325: 3–25, 129. [2018-12-01]. http://pubs.usgs.gov/circ/

Hungr O, Clague J J, Morgenstern N, et al. 2016. A review of landslide risk acceptability practices in various countries// Landslides and Engineered Slopes. Experience, Theory and Practice: 1121–1128.

Ionita I, Fullen M A, Zgłobicki W, et al. 2015. Gully erosion as a natural and human-induced hazard. Natural Hazards, 79(1): S1–S5.

Iverson R M, Reid M E, Iverson N R, et al. 2000. Acute Sensitivity of landslide rates to initial soil porosity. Science, 290(5491): 513–516.

Jaiswal P, van Westen C J, Jetten V. 2011. Quantitative assessment of landslide hazard along transportation lines using historical records. Landslides, 8(3): 279–291.

Jeong S, Kim B, Won J, et al. 2003. Uncoupled analysis of stabilizing piles in weathered slopes. Computers and Geotechnics, 30(8): 671–682.

Katz O, Morgan J K, Aharonov E, et al. 2014. Controls on the size and geometry of landslides: Insights from discrete element numerical simulations. Geomorphology, 220(3): 104–113.

Krohn K, Jaumann R, Otto K, et al. 2014. Mass movement on Vesta at steep scarps and crater rims. Icarus, 244: 120–132.

Lee C Y, Hull T S, Poulos H G. 1995. Simplified pile-slope stability analysis. Computers and Geotechnics, 17(1): 1–16.

Leshchinsky D, Baker R, Silver M L. 1985. Three-dimensional analysis of slope stability. International Journal for Numerical and Analytical Methods in Geomechanics, 9(3): 199–223.

Li Y, Shao M A. 2006. Effects of rainfall intensity on rainfall infiltration and redistribution in soil on Loess slope land. Chinese Journal of Applied Ecology, 17(12): 2271–2276 (in Chinese).

Lin G W, Chen H. 2012. The relationship of rainfall energy with landslides and sediment delivery. Engineering Geology, 125(1): 108–118.

Lohnes R A. 1991. A method for estimating land loss associated with stream channel degradation. Engineering Geology, 31(2): 115–130.

Lu N, Godt J W. 2013. Hillslope hydrology and stability. Cambridge: Cambridge University Press: 8–343.

Lucas A, Mangeney A, Mège D, et al. 2011. Influence of the scar geometry on landslide dynamics and deposits: Application to Martian landslides. Journal of Geophysical Research Planets, 116(E10): 1–21.

Luckman P G, Gibson R D, Derose R C. 1999. Landslide erosion risk to New Zealand pastoral steeplands productivity. Land Degradation and Development, 10(1): 49–65.

Meng Q M (Ed). 1996. Soil and water conservation on the Loess Plateau. Zhengzhou: Water Resource Press of Yellow River: 75 (in Chinese).

Millar R G, Quick M C. 1997. Discussion and closure: development and testing of riverbank-stability analysis, by Darby S E and Thorne C R J. American Society of Civil Engineers, 123(11): 1051–1053.

Morgenstern N R, Price V E. 1965. The analysis of the stability of general slip surfaces. Géotechnique, 15(1): 79–93.

Nearing M A, Pruski F F, O'neal M R. 2004. Expected climate change impacts on soil erosion rates: a review. Journal of Soil and Water Conservation, 59(1): 43–50.

Osman A M, Thorne C R. 1988. Riverbank stability analysis. I: Theory. Journal of Hydraulic Engineering, 114(2): 134–150.

Poesen J, Nachtergaele J, Verstraeten G, et al. 2003. Gully erosion and environmental change: importance and research needs. Catena, 50(2–4): 91-133.

Qiu H J, Cui P, Regmi A D, et al. 2017. Slope height and slope gradient controls on the loess slide size within different slip surfaces. Physical Geography, 38(4): 303–317.

Sayed S A, González P A. 2014. Flood disaster profile of Pakistan: A review. Science Journal of Public Health, 2(3): 144–149.

Shi J S, Wu L Z, Wu S R, et al. 2016. Analysis of the causes of large-scale loess landslides in Baoji, China. Geomorphology, 264(1): 109–117.

Simon A, Wolfe W J, Molinas A. 1991. Mass wasting algorithms in an alluvial channel model. Proceedings of the 5th Federal Interagency Sedimentation Conference, Las Vegas: 8–22.

Skepton A W, Hutchinson J N. 1969. Stability of natural slopes and embankment foundations. Proceedings of the 7th International Conference on Soil Mechanics and Foundation Engineering, Mexico: 291–340.

Stark T D, Choi H, McCone S. 2005. Drained shear strength parameters for analysis of landslides. Journal of Geotechnical and Geoenvironmental Engineering, 131(5): 575–588.

Tsai Z X, You G J Y, Lee H Y, et al. 2013. Modeling the sediment yield from landslides in the Shihmen Reservoir watershed, Taiwan. Earth Surface Processes and Landforms, 38(7): 661–674.

Waldmann N, Anselmetti F S, Ariztegui D, et al. 2011. Holocene mass-wasting events in Lago Fagnano, Tierra del Fuego (54° S): implications for paleoseismicity of the Magallanes-Fagnano transform fault. Basin Research, 23(2): 171–190.

Wang G Q, Li T J. 2009. The dynamic model of soil erosion and sediment transport in river basins. Beijing: China Water Power Press: 65 (in Chinese).

West A J, Hetzel R, Li G, et al. 2014. Dilution of 10Be in detrital quartz by earthquake-induced landslides: implications for determining denudation rates and potential to provide insights into landslide sediment dynamics. Earth and Planetary Science Letters, 396(15): 143–153.

Wu L Z, Zhou Y, Sun P, et al. 2017. Laboratory characterization of rainfall-induced loess slope failure. Catena, 150: 1–8.

Xu B, Low B K. 2006. Probabilistic stability analyses of embankments based on finite-element method. Journal of Geotechnical and Geoenvironmental Engineering, 132(11): 1444–1454.

Xu L, Dai F C, Tu X B, et al. 2013. Occurrence of landsliding on slopes where flow sliding had previously occurred: an investigation in a loess platform, North-west China. Catena, 104(5): 195–209.

Xu X Z, Liu Z Y, Xiao P Q, et al. 2015a. Gravity erosion on the steep loess slope: Behavior, trigger and sensitivity. Catena, 135: 231–239.

Xu X Z, Liu Z Y, Wang W L, et al. 2015b. Which is more hazardous: avalanche, landslide, or mudslide? Natural Hazards, 76(3): 1939–1945.

Yin Y P, Wang F W, Sun P. 2009. Landslide hazards triggered by the 2008 Wenchuan earthquake, Sichuan, China. Landslides, 6(2): 139–152.

Yitbarek T W, Belliethathan S, Stringer L C. 2012. The onsite cost of gully erosion and cost-benefit of gully rehabilitation: A case study in Ethiopia. Land Degradation and Development, 23(2): 157–166.

Zhang B L, Fan W. 2015. The effect on the stress distribution of the slope for rock and soil characteristics. Advanced Materials Research, 1065–1069: 227-231.

Zhang J Y, Gu P P, Li L Y, et al. 2016. Changes of soil particle size fraction along a chronosequence in sandy desertified land: a fundamental process for ecosystem succession and ecological restoration. Journal of Soils and Sediments, 16(12): 2651–2656.

Zhang M, Liu J. 2010. Controlling factors of loess landslides in western China. Environmental Earth Sciences, 59(8): 1671–1680.

Zhang Z P, Qu Y Z, Li H. 1997. Based on location and shape of landslide sliding surface to determine the engineering countermeasures. Subgrade Engineering, (5): 8–11 (in Chinese).

Zhou J, Fu B J, Gao G Y, et al. 2016. Effects of precipitation and restoration vegetation on soil erosion in a semi-arid environment in the Loess Plateau, China. Catena, 137: 1–11.

Zhou P H, Wang Z L. 1992. A study on rainstorm causing soil erosion in the Loess Plateau. Journal of Soil and Water Conservation, 6(3): 1–5 (in Chinese).

Zhu L, Hong B N. 2011. Sensitivity analysis on influencing factors of shallow landslide of coal measure strata. Journal of Yangtze River Scientific Research Institute, 28(7): 67–71,75 (in Chinese).

Zhuang J Q, Peng J B. 2014. A coupled slope cutting—a prolonged rainfall-induced loess landslide: a 17 October 2011 case study. Bulletin of Engineering Geology and the Environment, 73(4): 997–1011.

Zieliński A, Mazurkiewicz E, Łyskowski M, et al. 2016. Use of GPR method for investigation of the mass movements development on the basis of the landslide in Kałków. Roads and Bridges-DrogiiMosty, 15(1): 61–70.

Chapter 12
Effects of Gravity Erosion on Particle Size Distribution of Suspended Sediment

Abstract Gravity erosion generates an enormous volume of sediments on steep slopes throughout the world, yet its effect on the particle size distribution of suspended sediments (PSDSS) remains poorly understood. In this chapter, experiments were conducted in a field mobile laboratory in which mass movements were triggered on steep slopes under simulated rainfall. A suite of indexes such as median sediment size (d_{50}), sediment heterogeneity (H), fractal dimension (D), and enrichment/dilution ratio (R_{ed}) were used to evaluate the effect of mass movement on PSDSS. Mass movements led to a drastic increase in the sediment concentration and the enrichment of fine particles, which developed into hyperconcentrated flows. R_{ed}s for clay, silt, and sand fractions were 13.9, 1.4, and 0.7 respectively. The d_{50}, H, and D were significantly correlated with slope failures. The changes of PSDSS after mass movements reflected a combined complex effect of soil sources, erosion types, selective detachment, and deposition processes.

Keywords Gravity erosion · Landforms · Particle size distribution · Sediment · Soils

12.1 A Review on the Study of Particle Size Distribution of Suspended Sediment on Steep Slopes

Gravity erosion, also referred to as mass movement or mass wasting, is slope failures on steep slopes driven by self-weight, which is in contrast to other types of soil erosion processes that require physical impetus of wind or water. Forms of gravity erosion include landslide, avalanche, and earthflow (Xu et al. 2015a). Slope failures are not only a hazard in almost all countries, causing billions of dollars of damages and many casualties, but also contribute to landscape evolution and erosion in mountainous regions (Keefer and Larsen 2007). Moreover, slope failures could deliver a huge amount of sediments to rivers throughout the world, and seriously affect the structure and function of ecosystems and society. For example, a streamside landslide delivered 60,000 of mixed-size sediment to the Navarro River in California in the spring of 1995 (Sutherland et al. 2002). In some catchments of the Loess Plateau, mass movement contributed over 50% of the total sediment discharge

© Science Press and Springer Nature Singapore Pte Ltd. 2020   191
X. Xu et al., *Experimental Erosion*,
https://doi.org/10.1007/978-981-15-3801-8_12

(Xu et al. 2017). In the headwaters of the 2150 km^2 Waipaoa River Catchment, an area well known for its severe erosion in New Zealand, the amount of mass movement accounted for 41% of the total sediment load (Marden et al. 2014).

Particle size is one of the important characteristics of suspended sediment, because it can reflect the sediment source and erosion processes (Wendling et al. 2016; Woodward and Walling 2007; Xu 2002), affect the entrainment, transport and deposition of soils (Walling et al. 2000; Haritashya et al. 2010), and can be used to infer the contaminant sources (Slattery and Burt 1997; Smith and Owens 2014; Abarca et al. 2016). Sediment usually carries the signature of upstream disturbances in runoff and erosion to downstream channels (Sutherland et al. 2002). Previous studies used particle size characteristics as a means to trace suspended sediment in the river systems (Walling and Woodward 1995). Walling and Moorehead (1989) found that considerable variation existed in the particle size characteristics of sediment from different rivers in response to variations in source material and other physiographic controls. Jia et al. (2016) also demonstrated that particle size characteristics were useful in determining sediment provenance in the Yellow River basin based on the sediment deposits in the river system which mainly comprised coarse sand particles larger than 0.05 mm. Soil erosion causes sediments to be detached from their source materials and transported as suspended particles (Sadeghi and Harchegani 2012). In fact, soil erosion is a size-selective process. Fine particle-size fractions are selectively removed during inter-rill erosion process (Asadi et al. 2011; Wang et al. 2008; Stone and Walling 1997; Govers 1985), resulting in generally finer sediments compared to the source soil (Issa et al. 2006; Hao et al. 2016). Rill flows transport a greater proportion of larger particles as compared with interrill flows (Alberts et al. 1980).

Research on particle size distribution caused by mass movements has been limited and almost exclusively been done in the field after their occurrence. Crosta et al. (2007) and Davies and McSaveney (2009) reported that a large amount of small-size debris was generated by fragmentation and collision of the large-size debris of landslides. Cochrane and Acharya (2011) found that the largest particle in the runoff affected by shallow landslides was significantly smaller than that of the original material. However, it is unknown how particle size distribution changes during rainfall events in which both gravity erosion and water erosion occur. Such process-based data are hard, if not impossible, to obtain under natural rainfall conditions due to the unpredictable nature of gravity erosion in timing and location. Thus, experimental studies have become increasingly important in understanding the processes and mechanisms of gravity erosion. Chorley (1964) identified three broad classes of physical models in experiments, namely, segments of unscaled reality, scale models, and analog models. The monitoring of segments of unscaled reality is a widely used experimental approach, and it has a long history and has arrived at a good understanding of surficial erosion processes on hill slopes (Schumm et al. 1987). Xu et al. (2015b), for example, conducted a series of experiments in the laboratory to test the stability of slopes under different slope geometries and rainfall conditions, and then performed a sensitivity analysis to quantitatively explore the triggering mechanisms of mass failure on the steep loess slopes. Nevertheless, with respect to particle selectivity, previous rainfall simulation experimental studies are mostly confined to rain splash, sheet flow and rill flow (Hao et al. 2016; Govers 1985; Alberts et al. 1980). No experimental study has been conducted on the changes in the particle size distribution of suspended sediment during the mass failure processes. In this study, we carried out a series of experiments under simulated rainfall on the natural loess slopes on the Loess Plateau of China. The experiment is a closely monitored segment of unscaled reality. A suite of indexes such as median particle size (d_{50}), sediment heterogeneity (H), fractal dimension (D) and enrichment/dilution ratio (R_{ed}) are then used to evaluate the effect of gravity erosion on PSDSS. This study will be of theoretical and practical significance in understanding the sediment transport processes and hyper-concentrated flows, as well as tracing the sources of sediments in the Yellow River (Xu 2000).

12.2 Characteristics of the Liudaogou Catchment

Rainstorm-induced gravity erosion frequently occurs on the Loess Plateau due to the crisscrossing gullies of the undulating terrain, sparse vegetation, and numerous vertical joints in the loess deposits. Steep gully/valley banks with slope gradients more than 70° in the upper reaches of small watersheds are particularly susceptible to gravity erosion (Xu et al. 2015b). Sediment discharge from mass movement accounts for a considerable proportion of the total soil loss (Xu et al. 2017). The study site, Liudaogou Catchment (110°21′–110°23′ E, 38°46′–38°51′ N), is located in water-wind-gravity erosion crisscross region on the Loess Plateau, which is characterized by a large number of deep gullies and undulating loess slopes (Fig. 12.1a). In this area, gravity erosion, including avalanche, landslide, and earthflow, is very active on steep slopes, and contributes a large amount of sediment yield in the catchment. Field experimental results show that the amount of the gravity erosion might account for 67% of the total erosion on the steep slopes of the Liudaogou Catchment (Guo et al. 2016).

12.3 Methods and Materials

In the summer of 2014, a series of experiments were conducted on the natural loess slopes in the Liudaogou Catchment, Shenmu County (Fig. 12.1). Three conceptual landforms were "cut" in the field without disturbing the slope beneath. In other words, the original texture and density of soils on the landforms were kept unaltered, although the surface was cut to be smooth. Each of them covered an area of 3.0 m by 2.8 m. All landforms had a height of 1.5 m and a gentle upper slope of 3°. The steep lower slopes of the landforms were 60°, 70° and 80°, respectively. Underlying the surfaces was sandy loess, with the d_{50} of 0.108 mm. A mobile lab was set up in the experimental spot to keep out sunshine and winds. Each simulated rainfall had an intensity of 0.8 mm/min and a duration of 60 min. Five rainfalls were applied

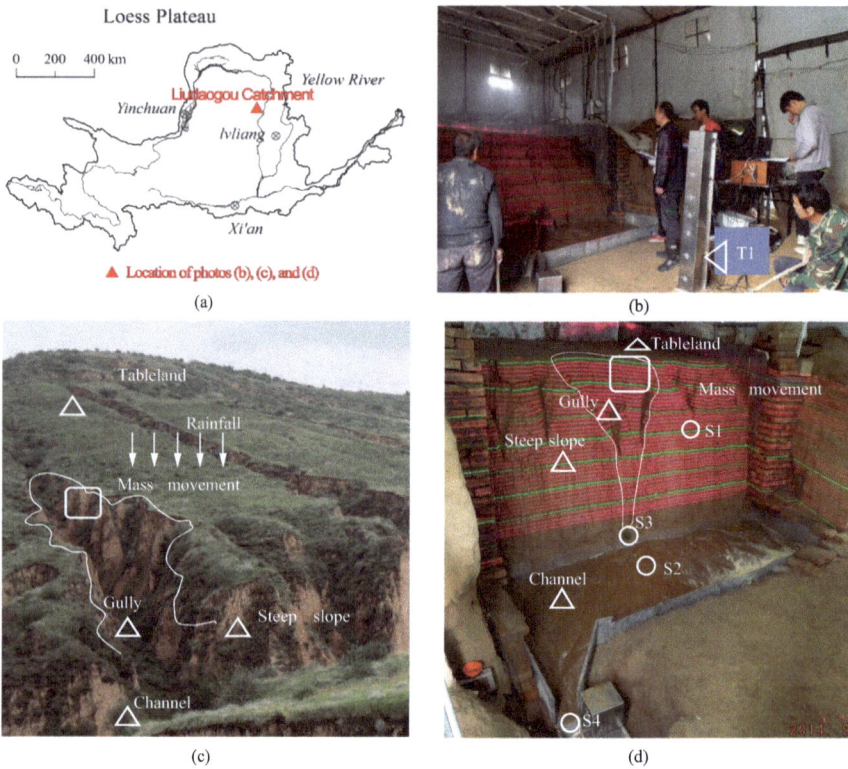

Fig. 12.1 Study area and sampling sites. **a** The location of the Liudaogou Catchment on the Loess Plateau of China; **b** A mass movement experiment in the Liudaogou Catchment, Shenmu County; **c** A photograph showing the typical mass movement topography on the gully heads; **d** Sampling sites in the experiment. T1: Topography meter, S1: undisturbed soil samples collected in the depth of 0–5 cm of the steep slope before the first event of rainfall, S2: deposited sediment samples collected from the channel bed after five rainfalls, S3: suspended sediment samples at the gully outlet, and S4: suspended sediment samples at the channel outlet

to each landform successively with a 12-h or so inter-rainfall interval to ensure the approximately similar antecedent water content. A novel structural laser-based topography meter was used to observe the mass failures in the experiments (Guo et al. 2016). With the topography meter, a 3D geometric shape of the target surface was digitally reconstructed, and then, the volume of mass failure on the dynamically changing steep slope was calculated.

Gullies emerged on the steep slope at the third rainfall on every experimental model landform. Normally two or three mass movements occurred on the gully head during the fourth and fifth rainfall. In order to obtain information on the particle size characteristics of suspended sediment, water samples were collected in the pre-cleaned bottles (500 ml) directly from the gully and channel flows. Here, GS denote the suspended sediment samples when the mass failures occurred on the gully head, and WS denote those that were not disturbed by mass failures when water erosion occurred only. In addition, on each model landform, before the first event of rainfall, soil samples were taken in the depth of 0–5, 40–45 and 120–125 cm from the steep slope section. After the fifth rainfall, samples were collected from the deposited sediments on the channel bed. In the experiment, the deposited sediments on the channel bed were cleared away after each rainfall event, so their thickness was only about 10 cm at the end of the fifth rainfall. Suspended sediment samples were collected at the gully and channel outlets, and the deposited sediment samples were collected in the middle of the channel bed (Fig. 12.1d). Of all the samples collected, 8 were gravity erosion samples (GS), 6 were water erosion samples (WS), 6 were undisturbed soil samples (US), and 2 were deposited sediment samples (SD). The locations of the sampling sites and their characteristics are also listed in Table 12.1.

All of the suspended sediment samples were dried, and then the distribution of the particles with sizes larger than 0.1 mm was determined by the sieving method, and the distribution of those with sizes less than 0.1 mm fractions were classified by the photoelectric sedimentometer technique. According to the taxonomy of the United States Department of Agriculture, the PSDSS were separated into three size fractions: <0.002 mm (clay), 0.002–0.05 mm (silt), and 0.05–2.0 mm (sand). The mass percentage of each particle size fraction was the weight of each fraction divided by the sum of all particle fractions.

The sediment heterogeneity H was calculated by the ratio of d_{60} to d_{10} (Olsen and Townsend 2003):

$$H = d_{60}/d_{10} \tag{12.1}$$

where d_{60} and d_{10} are the particle diameters at 60 and 10% in the cumulative distribution, respectively.

Table 12.1 Summary information for the soil sampling sites

Sampling site	Sample type	Sample no.	Sampling time
S1: steep slope	US	6	Before the first rainfall
S2: channel bed	SD	2	After the fifth rainfall
S3: gully outlet	LS	2	During the fourth rainfall and fifth rainfall
	ES	2	
	WS	4	
S4: channel outlet	LS	2	During the fourth rainfall and fifth rainfall
	ES	2	
	WS	2	

Notes (1) Sampling sites are shown in Fig. 12.1d. (2) LS, ES denote the suspended sediment samples when landslide and earthflow occurred, respectively; WS denote the suspended sediment samples were not disturbed by mass failures as water erosion occurred only; US and SD represent the samples of the undisturbed soil on the slope and the deposited sediment at the channel bed, respectively

The fractal Dimension (D) of the suspended sediment was estimated as follows (Tyler and Wheatcraft 1992):

$$D = 3 - \frac{\lg\left[\frac{M(r<R_i)}{M_T}\right]}{\lg\left(\frac{R_i}{R_{\max}}\right)} \qquad (12.2)$$

where $M(r < R_i)$ is the cumulative mass percentage of the particles of the ith size for r less than R_i, M_T is the total percentage ($M_T = 100$), R_i is the mean particle radius (mm) of the ith size class, and $R_{\max}$ is the radius of the largest particle class. $R_{\max}$ is 2.0 mm in this study. Equation (12.2) has been widely used to determine the D value of the PSDSS (Lyu et al. 2015; Wang et al. 2010).The enrichment/dilution ratio, $R_{ed,}$ was calculated as follows:

$$R_{ed} = S_s/S_b \qquad (12.3)$$

where S_s is the weight percentage of the particles in a given size class in the runoff sample, and S_b is that in the source soil. R_{ed} values greater than 1.0 represent enrichment: a given size class forms a greater proportion of the transported load in the runoff than in the source soil. The larger the R_{ed} is, the stronger the enrichment becomes. The value of R_{ed} could clearly reflect the change of different groups of soil particles.

12.4 Results and Discussion

12.4.1 Changes of PSDSS Affected by Gravity Erosion

The PSDSS for the samples from the gully and channel flows is given in Table 12.2 and Fig. 12.2. Prior to the gravity erosion, the sand-, silt- and clay-sized fractions of suspended sediments in gully flows averaged 71, 28 and 1%, respectively. It was noted that the numbers in parenthesis represented the range of percentage in a given particle size fraction. After mass failures occurred, the sand sized fraction decreased to a mean value of 51% whereas silt- and clay-sized particle fraction increased to 42 and 7%, respectively. Similar trends were observed in channel flows. Sand-sized fraction was decreased from an average of 64–44%, whereas silt-particle fractions increased from 31 to 51%, and clay-sized particle fraction remained unchanged after mass failures. There were also significant differences in PSDSS among different types of mass failures. Earthflows produced a lower percentage of sand-sized sediment in gully flows, but a higher one in channel flows than landslides did (Fig. 12.2).

A further comparison of the suspended sediments in gully flows with the undisturbed surface soils indicated that both were very similar in particle size distribution prior to mass failures, with the mean value of sand-, silt- and clay-fraction of

Table 12.2 PSDSS in the gully and channel flow

Sampling site	Sample no.	PSDSS (%)			Sampling site	Sample no.	PSDSS (%)		
		Clay <0.002 mm	Silt 0.002–0.05 mm	Sand 0.05–2 mm			Clay <0.002 mm	Silt 0.002–0.05 mm	Sand 0.05–2 mm
Gully (S3)	WS 1	1	16	83	Channel (S4)	WS 5	5	33	61
	WS 2	1	22	77		WS 6	5	29	66
	WS 3	2	35	63		–	–	–	–
	WS 4	1	38	61		–	–	–	–
	Mean	1	28	71		Mean	5	31	64
	LS 1	5	36	59		LS 3	6	61	33
	LS 2	5	36	59		LS 4	5	60	35
	ES 1	10	50	40		ES 3	4	42	54
	ES 2	10	45	45		ES 4	4	40	56
	Mean	7	42	51		Mean	5	51	44

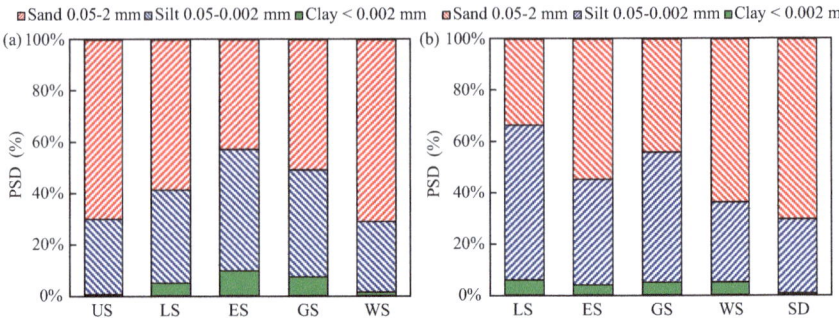

Fig. 12.2 Particle size distribution of the suspended sediment, slope soil and deposited sediment. Sampling sites S1–S4 are shown in Fig. 12.1d and Table 12.1. **a** Samples from sites S1 and S3; **b** Samples from sites S2 and S4. GS denote the suspended sediment samples during the mass failures, including landslide and earthflow

71, 28 and 1% versus 70, 29 and 1%, respectively (Fig. 12.2a). However, the sand-sized particle fraction in gully flows was significantly smaller than that of surface soils after gravity erosion occurred. Regardless of gravity erosion, channel flows contained less sand-sized particles than both surface soils and deposited sediments on channel bed (Fig. 12.2b).

12.4.2 Effects of Gravity Erosion on d_{50}, H, D

To further quantitatively assess the effect of gravity erosion on PSDSS, a suite of particle size indexes were used. Figure 12.3a shows a significant variation in the median particle size of the suspended sediments. The d_{50} of the gravity erosion samples was significantly smaller than that of the water erosion samples from both gully and channel flows. The d_{50} was 0.084 mm in the gully flows and 0.063 mm in the channel flows prior to the gravity erosion, and was decreased to 0.051 mm in the gully flows and 0.043 mm in the channel flows as gravity erosion occurred. That is to say, the suspended sediment becomes finer with gravity erosion. The values of d_{50} in this study are within the range of the PSDSS in the Yellow River (Xu 2000).

Sediment heterogeneity (H) is a non-uniform index. Prior to the gravity erosion, H was very small for the gully flows with a mean value of 5.6, which was similar to that of the undisturbed soil on the slope surface with a mean value of 4.3. However, it was remarkably increased to 26.8 as gravity erosion occurred (Fig. 12.3b). It was clear that gravity erosion led the suspended sediment to more heterogeneous and the sediments were transported in a more complicated manner (Holland and Elmore 2008).

The fractal dimension (D) is a measure of complex form irregularity. D can not only compare the characteristics of grain distribution and the uniformity of particle texture, but also describe the complex and irregular PSDSS. All D values in this

Fig. 12.3 Variation of the suspended sediment properties during the rainfall. **a–c** shows the changes of the median particle size, sediment heterogeneity, and fractal dimension, respectively

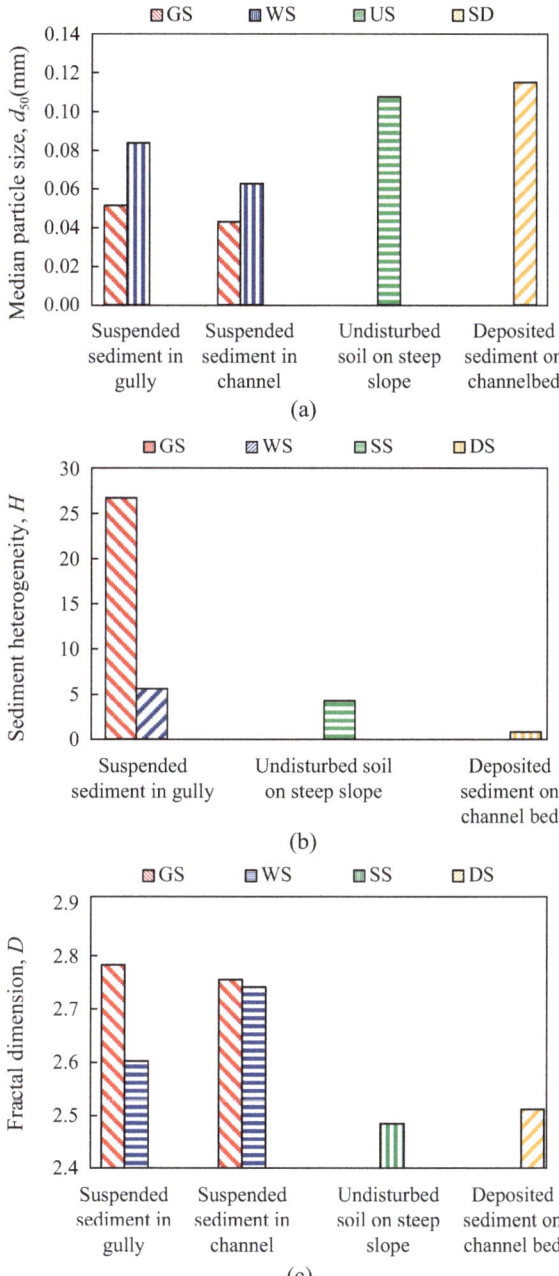

study were within the expected range from 0 to 3 (Turcotte 1986). The D values of the suspended sediments ranged from 2.60 to 2.78, which were all higher than those of the undisturbed slope soils, with an average value of 2.48. As shown in Fig. 12.3c, there was a significant difference in D among different types of soil erosion. D increased from 2.60 to 2.78 for the gully flows as gravity erosion occurred. This implies that mass failures could make the PSDSS more irregular, because the proportions of silt and clay particles were increased and the proportion of sand-sized particles was decreased, which led to an increase in the D values (Gao et al. 2014; Wang et al. 2010). Nevertheless, the D values of the suspended sediments in the channel flows before and after mass failures were quite close (2.74 vs. 2.76). The results demonstrate that D could be one of the effective indexes to reflect the physical properties of the suspended sediment by different soil erosion processes.

12.4.3 Changes of Enrichment/Dilution Ratios by Water Erosion and Gravity Erosion

Enrichment ratio (R_{ed}) compares the PSDSS with the undisturbed source soils. It could clearly reflect the degree of change in the particle size fractions after erosion occurs (Zhang et al. 2016; Slattery and Burt 1997; Walling and Moorehead 1989). Figure 12.4 presented the summary information on the R_{ed} of the suspended sediment samples. For all samples, especially gravity erosion samples, the suspended sediment tended to be enriched in the silt and clay fractions and diluted in the sand fractions. For gravity erosion samples in the gully flows, R_{ed} was at the highest value of 13.9 for clay fraction, 1.4 for silt fraction and at the lowest value 0.7 for sand. R_{ed}s of the clay, silt and sand fractions were 2.5, 0.9 and 1.0, respectively, in the gully flows when water erosion occurred only. For gravity erosion samples in the channel flows, R_{ed}

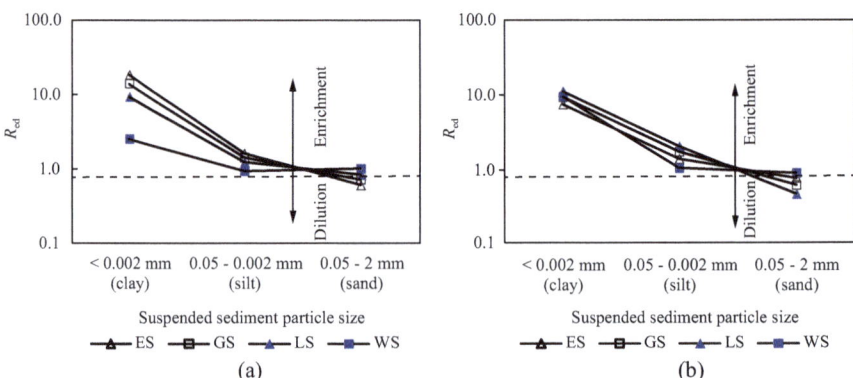

Fig. 12.4 Enrichment/dilution ratios of the PSDSS. **a** Suspended sediment in the gully runoff; **b** Suspended sediment in the channel flow

was at the highest value of 9.3 for clay fraction, 1.7 for silt fraction and at the lowest value 0.6 for sand fraction. $R_{ed}s$ of the clay, silt and sand fractions in channel flows were 9.3, 1.4 and 0.9, respectively, when water erosion occurred only. It can be seen that the change of $R_{ed}s$ of the suspended sediment particle sizes caused by gravity erosion and water erosion in the channel flow followed the similar trends to that in the gully flows. However, the difference of $R_{ed}s$ in the channel flows influenced by gravity erosion and water erosion became smaller than in the gully flows.

12.4.4 Causes of the Changes of PSDSS

The changes of PSDSS before and after mass failures reflected the combined effects of soil sources, erosion types, as well as sediment transport and depositional processes. The soil type and characteristics were the dominant factors controlling the particle size (Wendling et al. 2016). Soil textures at each location determine the range and availability of particle sizes for possible erosion (Grismer et al. 2008). In addition, disaggregation of unstable aggregates due to the impact of raindrops and runoff turbulence can also affect the particle size distribution of the eroded sediment (Meyer et al. 1992). Moreover runoff velocity might also have an effect on the sediment particle size distribution to a certain extent (He et al. 2017).

Prior to the mass failures, the four suspended sediments samples taken from gully flows showed a graduate decrease in sand-sized fraction over time, in the order of 83, 77, 63 and 61% (Table 12.2). The initial high contents of sand-sized particles might be the loose and large particles on the surfaces which were washed away by the overland flow. With the depletion of those large and loose particles, the detachment of cohesive soil particles required more energy and therefor might be size-selective, which led to the decrease of sand-sized fraction and the enrichment of fine particles. These results are consistent with other studies that show the dynamic changes of particle size selectivity in water erosion processes (Koiter et al. 2017; Shi et al. 2013; Issa et al. 2006).

The smaller fraction of sand-sized particles in channel flows than that in gully flows, with a mean value of 64 versus 71%, indicates the selective deposition of coarse sediments on channel beds due to the abrupt decrease of slope gradient from gully slope to channel bed. It is expected no or very little deposition occurred in the steep gullies. After mass failures, the fraction of sand-sized particles in gully flows further decreased, with a mean value of 51% ranging from 40 to 59%. Such a decrease was likely to be caused by the source materials from mass movements and the selective detachment by overland flows. Mass failures occurred within a thicker soil layer with lower content of sand-sized particles than that on surface soils. As shown in Table 12.3, the average proportion of sand-sized particle was 70% on surface soils (0–5 cm), in comparison of 60% on the depth of 40–45 cm and 59% on the depth of 120–125 cm. The median particle size also decreased with the depth (Fig. 12.5), which may be ascribed to the different climatic conditions during the deposition of wind-blown loess particles in Quaternary and Holocene. The fraction

Table 12.3 PSDSS for various depths

Sampling depth (cm)	Sample no.	PSDSS (%)			d_{50} (mm)
		Clay	Silt	Sand	
		<0.002 mm	0.002–0.05 mm	0.05–2 mm	
0–5	US1	1	31	68	0.105
	US2	0	28	72	0.110
	Mean	1	29	70	0.108
40–45	US3	1	39	60	0.072
	US4	1	41	59	0.073
	Mean	1	40	60	0.073
120–125	US5	1	45	54	0.055
	US6	1	35	64	0.062
	Mean	1	40	59	0.059

Fig. 12.5 The median particle size of soil (d_{50}) versus soil depth

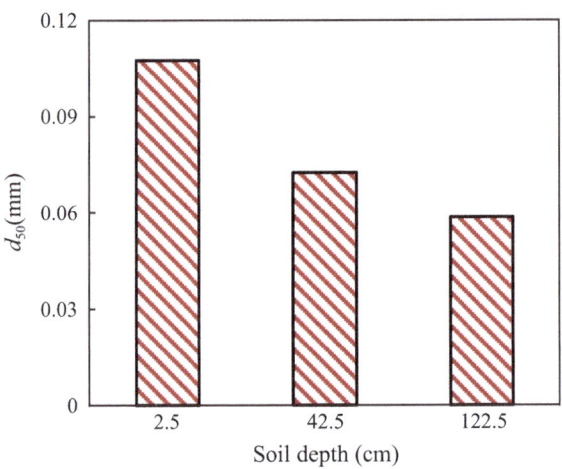

of sand-sized particles in channel flows also showed a further decrease to a mean value of 44% after mass failures in the comparison of 61% prior to mass failures. Nevertheless, the deposited sediments on the channel bed had a sand content of 70%, which was a clear indication of the selective deposition of coarse particles after mass failures. Such selective accumulation of coarse materials on the channel beds can form armouring surfaces in the field.

This study demonstrated the combined complex effects of source materials, erosion types, and selective detachment and deposition processes on PSDSS before and after mass failures, though some of these individual factors have been investigated by others (Asadi et al. 2011; Beuselinck et al. 2002; Collin et al. 1997; Gover 1985). Our experimental results in PSDSS also compare well with the previous field studies in the region (e.g., Wang et al. 2010). Wendling et al. (2016) indicated that the suspended concentration controls the particle size distribution to some extent. Increasing the suspended concentration led to more frequent collisions between particles (Abrahamson 1975). If the probability of breakdown or abrasion during a collision is higher than the probability of cohesion, the statistical effect of an increasing collision number is a decreased mean particle size.

12.4.5 Gravity Erosion and Hyper-Concentrated Flows

A considerable number of field observations indicated that hyperconcentrated flows may be associated with mass movements (Li et al. 2009; Pierson 2005; Xu 1999). There are different definitions of hyper-concentrated flows. Qian (1989) refers them to the flows with sediment concentration greater than 800 kg/m^3. In this study, we experimentally observed a drastic increase in the sediment concentration after mass failures, generating hyper-concentrated flow. As shown in Table 12.4 and Fig. 12.6, sediment concentrations in the gully flows ranged from 612.0 to 763.5 kg/m^3 when

Table 12.4 Correlation between the sediment particle size and concentration

Sampling site		Sample no.	PSDSS (%)		Sediment concentration (kg/m^3)
			<0.05 mm	0.05–2 mm	
Gully runoff	Water erosion	WS1	17	83	612.0
		WS2	23	77	637.8
		WS3	37	63	670.4
		WS4	39	61	763.5
		Mean	29	71	670.9
	Gravity erosion	LS1	41	59	1005.4
		LS2	41	59	1145.0
		ES1	60	40	1001.3
		ES2	55	45	994.1
		Mean	49	51	1036.4
Channel flow	Water erosion	WS5	39	61	8.8
		WS6	34	66	6.6
		Mean	36	64	7.7
	Gravity erosion	LS3	68	32	37.8
		LS4	65	35	31.5
		ES3	46	54	36.7
		ES4	44	56	21.8
		Mean	56	44	32.0

Fig. 12.6 The clay and silt content versus sediment concentration

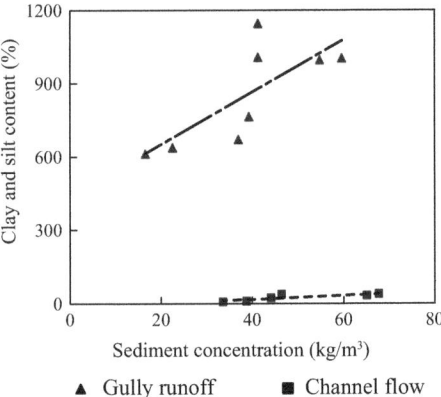

water erosion occurred only. Nevertheless, they were increased to the range of 994.1–1145.0 kg/m³ after slope failures. Furthermore, with the increase of the sediment concentration, the clay and silt content of the gully runoff grew more rapidly than in the channel (Fig. 12.6). High sediment contents were found during slope failures by others in the field studies. For examples, Acharya et al. (2011) observed that the peak sediment yields at the failure time were in the range 10–50 times higher than the post-failure yields. Ries (2000) reported a three-day-long landslide event in a river basin of Central Nepal, in which the landslide-driven sediment yield was twice the total annual sediment yield.

Besides the excessively high sediment concentrations in the hyperconcentrated flow after mass failures, particle sizes also varied. The clay and silt content (<0.05) was increased from 29 to 49% when sediment concentration rose from 670.9 to 1036.4 kg/m³ in the gully flows (Table 12.4). In contrast, the sand content (0.05–2 mm) was decreased with the rising sediment concentrations (Table 12.4).

The mechanism of hyperconcentrated flow is not fully understood. As a turbulent subaerial flow, the hyperconcentrated flow is excessively dense and hence it deposits sediment mainly or entirely in a non-traditional manner. Shanmugam (1996) explains that the sediment-support mechanisms in hyperconcentrated flows may include matrix strength, dispersive pressure, and buoyant lift. Xu (1999) considers the hyperconcentrated flow as a turbulent flow with the two phases of solid and liquid. A large amount of fine sediment uniformly mixed with water forms the liquid phase, in which the relatively coarse particles are suspended as the solid phase. Due to a large unit weight of turbid water, the submerged weight of coarse particles becomes relatively smaller, thus their settling velocity is also smaller, and so is the work required to be expended for their suspension. In the present study, both the great number of sediments and the enrichment of fine particles resulting from gravity erosion favor the generation of hyperconcentrated flows. It would be interesting to determine what is the maximum ratio of coarse to fine sediments in the hyperconcentrated flows and how this ratio changes with the sediment concentration, which we are going to study in the future.

12.5 Conclusions

PSDSS showed a great difference before and after mass failures. As the gravity erosion occurred, the proportion of sand-sized particles was decreased from 71 to 51%, whereas the proportions of clay and silt were increased remarkably from 1 to 7% and 28 to 42%, respectively. Besides, the PSDSS had a significant difference among different gravitational erosion types. The d_{50}, H and D were significantly correlated with gravity erosion. As a result of gravity erosion, d_{50} was decreased from 0.084 to 0.051 mm, H was increased from 5.6 to 26.8, and D was increased from 2.60 to 2.78. The results imply that gravity erosion made the PSDSS more non-uniform and irregular. Suspended sediment tended to enrich in the silt and clay fractions and dilute in the sand fractions during mass movement, with R_{ed}s of 13.9, 1.4 and 0.7 for clay, silt, and sand, respectively. The changes of PSDSS after mass failures reflected a combined complex effect of soil sources, erosion types, selective detachment and deposition processes. Mass failures led to the drastic increase of sediment concentration and the enrichment of fine particles, which developed hyperconcentrated flows.

References

Abarca M, Guerra P, Arce G, et al. 2016. Response of suspended sediment particle size distributions to changes in water chemistry at an Andean mountain stream confluence receiving arsenic rich acid drainage. Hydrological Processes, 31(2): 296–307.

Abrahamson J. 1975. Collision rates of small particles in a vigorously turbulent fluid. Chemical Engineering Science, 30(11): 1371–1379.

Acharya G, Cochrane T, Davies T, et al. 2011. Quantifying and modeling post-failure sediment yields from laboratory-scale soil erosion and shallow landslide experiments with silty loess. Geomorphology, 129(1–2): 49–58.

Alberts E E, Moldenhauer W C, Foster G R. 1980. Soil aggregates and primary particles transported in rill and interrill flow. Soil Science Society of America Journal, 44(3): 590–595.

Asadi H, Moussavi A, Ghadiri H, et al. 2011. Flow-driven soil erosion processes and the size selectivity of sediment. Journal of Hydrology, 406(1–2): 73–81.

Beuselinck L, Govers G, Hairsine P B, et al. 2002. The influence of rainfall on sediment transport by overland flow over areas of net deposition. Journal of Hydrology, 257(1–4): 145–163.

Chorley R J. 1964. Geography and analogue theory. Annals of the Association of American Geographers 54(1), 127–137.

Cochrane T A, Acharya G. 2011. Changes in sediment delivery from hillslopes affected by shallow landslides and soil armouring. Journal of Hydrology, 50(1): 5–18.

Collins A L, Walling D E, Leeks G J L. 1997. Source type ascription for fluvial suspended sediment based on a quantitative composite fingerprinting technique. Catena, 29(1): 1–27.

Crosta G B, Frattini P, Fusi N. 2007. Fragmentation in the Val Pola rock avalanche, Italian Alps. Journal of Geophysical Research Earth Surface, 112:F01006.

Davies T R, McSaveney M J. 2009. The role of rock fragmentation in the motion of large landslides. Engineering Geology, 109(1–2): 67–79.

Gao G L, Ding G D, Wu B, et al. 2014. Fractal scaling of particle size distribution and relationships with topsoil properties affected by biological soil crusts. Plos One, 9(2): e88559.

Govers G. 1985. Selectivity and transport capacity of thin flows in relation to rill erosion. Catena, 12(1): 35–49.

Grismer M E, Ellis A L, Fristensky A. 2008. Runoff sediment particle sizes associated with soil erosion in the Lake Tahoe Basin, USA. Land Degradation and Development, 19(3): 331–350.

Guo W Z, Xu X Z, Wang W L, et al. 2016. A measurement system applicable for landslide experiments in the field. Review of Scientific Instruments, 87(4): 044501.

Hao Y F, Yang Y, Liu B Y, et al. 2016. Size characteristics of sediments eroded from three soils in China under natural rainfall. Journal of Soils and Sediments, 16(8): 2153–2165.

Haritashya U K, Kumar A, Singh P. 2010. Particle size characteristics of suspended sediment transported in meltwater from the Gangotri Glacier, central Himalaya - An indicator of subglacial sediment evacuation. Geomorphology, 122(1–2): 140–152.

He J J, Sun LY, Gong H L, et al. 2017. Laboratory studies on the influence of rainfall pattern on rill erosion and its runoff and sediment characteristics. Land Degradation and Development, 28(5): 1615–1625.

Holland K T, Elmore P A. 2008. A review of heterogeneous sediments in coastal environments. Earth-Science Reviews, 89(3–4): 116–134.

Issa O M, Bissonnais Y L, Planchon O, et al. 2006. Soil detachment and transport on field and laboratory scale interrill areas: erosion processes and the size-selectivity of eroded sediment. Earth Surface Processes and Landforms, 31(8): 929–939.

Jia X P, Li Y S, Wang H B. 2016. Bed sediment particle size characteristics and its sources implication in the desert reach of the Yellow River. Environmental Earth Sciences, 75(11): 950.

Keefer D K, Larsen M C. 2007. Assessing landslide hazards. Science, 316(5828): 1136–1138.

Koiter A J, Owens P N, Petticrew E L, et al. 2017. The role of soil surface properties on the particle size and carbon selectivity of interrill erosion in agricultural landscapes. Catena, 153: 194–206.

Li T J, Wang G Q, Xue H, et al. 2009. Soil erosion and sediment transport in the gullied Loess Plateau: scale effects and their mechanisms. Science in China Series E: Technological Sciences, 52(2): 1283–1292.

Lyu X F, Yu J B, Zhou M, et al. 2015. Changes of soil particle size distribution in tidal flats in the Yellow River Delta. Plos One, 10(3): e0121368.

Marden M, Herzig A, Basher L. 2014. Erosion process contribution to sediment yield before and after the establishment of exotic forest: Waipaoa catchment, New Zealand. Geomorphology, 226: 162–174.

Meyer L D, Line D E, Harmon W C. 1992. Size characteristics of sediment from agricultural soils. Journal of Soil and Water Conservation, 47(1): 107–111.

Olsen D A, Townsend C R. 2003. Hyporheic community composition in a gravel-bed stream: influence of vertical hydrological exchange, sediment structure and physicochemistry. Freshwater Biology, 48(8): 1363–1378.

Pierson T C. 2005. Hyperconcentrated flow-transitional process between water flow and debris flow. In Debris-flow hazards and related phenomena. Springer Berlin Heidelberg: 159–202.

Qian N. 1989. Movements of hyper-concentration flows. Beijing: Qinghua University Press (in Chinese).

Ries J B, 2000. The landslide in the Surma Khola valley, High Mountain Region of the Central Himalaya in Nepal. Physics and Chemistry of the Earth Part B Hydrology Oceans and Atmosphere, 25(1): 51–57.

Sadeghi S H, Harchegani M K. 2012. Effects of sand mining on suspended sediment particle size distribution in Kojour Forest River, Iran. Journal of Agricultural Science and Technology, 14(3): 1637–1646.

Schumm S A, Mosley M P, Weaver W E. 1987. Experimental fluvial geomorphology. New York: John Wiley and Sons: 1–7.

Shanmugam G. 1996. High-density turbidity currents: are they sandy debris flows? Journal of Sedimentary Research, 66(1): 2–10.

Shi Z H, Yue B J, Wang L, et al. 2013. Effects of mulch cover rate on interrill erosion processes and the size selectivity of eroded sediment on steep slopes. Soil Science Society of America Journal, 77(1): 257–267.

Slattery M C, Burt T P. 1997. Particle size characteristics of suspended sediment in hillslope runoff and stream flow. Earth Surface Processes and Landforms, 22(8): 705–719.

Smith T B, Owens P N. 2014. Flume- and field-based evaluation of a time-integrated suspended sediment sampler for the analysis of sediment properties. Earth Surface Processes and Landforms, 39(9): 1197–1207.

Stone P M, Walling D E. 1997. Particle size selectivity considerations in suspended sediment budget investigations. Water, Air and Soil Pollution, 99(1–4): 63–70.

Sutherland D G, Ball M H, Hilton S J, et al. 2002. Evolution of a landslide-induced sediment wave in the Navarro River, California. Geological Society of America Bulletin, 114(8): 1036–1048.

Turcotte D L. 1986. Fractals and fragmentation. Journal of Geophysical Research Solid Earth, 91(B2): 1921–1926.

Tyler S W, Wheatcraft S W. 1992. Fractal scaling of soil particle-size distributions: analysis and limitations. Soil Science Society of America Journal, 56(2): 362–369.

Walling D E, Moorehead P W. 1989. The particle size characteristics of fluvial suspended sediment: an overview. Hydrobiologia, 176(1): 125–149.

Walling D E, Woodward J C. 1995. Tracing sources of suspended sediment in river basins: a case study of the River Culm, Devon, UK. Marine and Freshwater Research, 46(1): 327–336.

Walling D E, Owens P N, Waterfall B D, et al. 2000. The particle size characteristics of fluvial suspended sediment in the Humber and Tweed catchments, UK. Science of the Total Environment, 251–252: 205–222.

Wang D, Fu B J, Zhao W W, et al. 2008. Multifractal characteristics of soil particle size distribution under different land-use types on the Loess Plateau, China. Catena, 72(1): 29–36.

Wang Y C, Liu K, Lau D L, et al. 2010. Maximum SNR pattern strategy for phase shifting methods in structured light illumination. Journal of the Optical Society of America A, 27(9): 1962–1971.

Wendling V, Legout C, Gratiot N, et al. 2016. Dynamics of soil aggregate size in turbulent flow: respective effect of soil type and suspended concentration. Catena, 141: 66–72.

Woodward J C, Walling D E. 2007. Composite suspended sediment particles in river systems: their incidence, dynamics and physical characteristics. Hydrological Processes, 21(26): 3601–3614.

Xu J X. 1999. Erosion caused by hyperconcentrated flow on the Loess Plateau of China. Catena, 36(1–2): 1–19.

Xu J X. 2000. Grain-size characteristics of suspended sediment in the Yellow River, China. Catena, 38(3): 243–263.

Xu J X. 2002. Implication of relationships among suspended sediment size, water discharge and suspended sediment concentration: the Yellow River basin, China. Catena, 49(4): 289–307.

Xu X Z, Liu Z Y, Wang W L, et al. 2015a. Which is more hazardous: avalanche, landslide, or mudslid? Natural Hazards, 76(3): 1939–1945.

Xu X Z, Liu Z Y, Xiao P Q, et al. 2015b. Gravity erosion on the steep loess slope: Behavior, trigger and sensitivity. Catena, 135: 231–239.

Xu X Z, Guo W Z, Liu Y K, et al. 2017. Landslides on the Loess Plateau of China: a latest statistics together with a close look. Natural Hazards, 86(3): 1393–1403.

Zhang J Y, Gu P P, Li L Y, et al. 2016. Changes of soil particle size fraction along a chronosequence in sandy desertified land: a fundamental process for ecosystem succession and ecological restoration. Journal of Soils and Sediments, 16(12): 2651–2656.

Chapter 13
Tunnel Flow and Erosion Processes in an Experimental Catchment

Abstract Soil pipes and tunnels have been reported in a wide range of climatic conditions in the world, yet their hydrological and sediment processes have been much understudied in comparison with surface flows. In this chapter, storm flows and sediment processes were monitored at 6 tunnel systems in an experimental catchment in the Loess Plateau. The results indicate that all the tunnel flow was derived from overland flow entering via inlets. However, both flow discharges and sediment concentrations were highly erratic due to the instability of tunnel systems. Tunnels could be clogged by collapse inside, which reopened in subsequent events. An intensive storm could develop new inlet(s), which either joined the existing tunnel system or started a new one. Partial damming within the tunnel systems also frequently occurred. This chapter showed that at least 43% of water flows and 57% of sediments were routed via tunnels in the experimental catchment.

Keywords Tunnel systems · Hydrological processes · Sediment processes · Tunnel instability · Contributions to catchment

13.1 A Brief Introduction on the Tunnel Erosion

Soil pipes or tunnels have been reported in many parts of the world, including Europe (Verachtert et al. 2010; Piccarreta et al. 2006; Holden 2005; Faulkner et al. 2004; Jones 1987; Harvey 1982), North America (Wilson et al. 2015; Seppälä 1997; Parker et al. 1990; Swanson et al. 1989; Drew 1982; Heede 1971; Parker and Jenne 1967), Asia (Zhu 2003; Onda 1994), Australia (Crouch ct al. 1986) and Africa (Frankl et al. 2016; Boucher and Powell 1994; Dardis and Beckedahl 1988; Yair et al. 1980). Soil piping/tunnel erosion has been increasingly recognized as a significant hillslope hydrological and geomorphologic process (Carey and Woo 2000; Uchida et al. 1999; Zhu 1997; Bryan and Jones 1997; Jones and Bryan 1997; Walsh and Howells 1988; Crouch et al. 1986; Bryan and Harvey 1985; Wilson and Smart 1984; McCaig 1983; Jones 1981; Gilman and Newson 1980). However, quantitative studies on pipe-flow and erosion processes are generally very limited and most of these studies have been conducted in humid areas (Walsh and Howells 1988; Jones 1987, Jones 1981; Jones and Crane 1984; McCaig 1983; Gilman and Newson 1980), and comparatively few

© Science Press and Springer Nature Singapore Pte Ltd. 2020
X. Xu et al., *Experimental Erosion*,
https://doi.org/10.1007/978-981-15-3801-8_13

studies have been conducted in semiarid or arid areas. Moreover, these field programs were often designed to monitor hydrological responses of pipe-flow to natural rainstorms. The pipe-flow erosion processes and their relationship to hydrological processes are still not fully understood. The relative significance of surface and subsurface erosion in hillslope sediment delivery, and the overall contributions of subsurface erosion to basin sediment yield are largely unknown in many parts of the world.

This chapter describes the field monitoring program conducted on the Loess Plateau (Zhu 1997, 2003, 2006, 2012; Zhu et al. 2000, 2002). Tunnel systems on the Loess Plateau of China are among the largest and most complicated ones in the world. Tunnel erosion and its controlling factors have been intermittently discussed by several people (Wang 1989; Zhu 1958; Thorp 1936; Fuller 1922), but mainly at qualitative level. Since 1989, an intensive field study has been conducted in Yangdaogou, a small catchment in western Shanxi Province. Sediment and runoff processes of tunnel flows have been monitored at the tunnel outlets over a two-year period and the tunnel systems have been repeatedly surveyed in the study catchment. The field study methods, hydrological and sediment processes of tunnel flows, significance of tunnel erosion in the catchment are discussed in the following sections.

13.2 Characteristics of the Wangjiagou Catchment

The study was conducted in the hilly-gully sub-region of the Loess Plateau. The field experiment site, Yangdaogou, is one of the sub-watershed in the Wangjiagou Catchment, with the area of 0.203 km^2 (Fig. 13.1). Wangjiagou is a long-term experimental watershed run by the Shanxi Institute of Soil and Water Conservation (SISWC). The study site is situated in the silty loess zone (Chen and Luk 1989). Vertically, loess consists of the upper Pleistocene Malan Loess, the Mid Pleistocene Lishi Loess, and the Lower Pleistocene Wucheng Loess (Liu 1964).

The climate of the study site is semiarid temperate with a mean annual precipitation of 500 mm or so, among which over 70% falls in the summer monsoon season (June to September). Both overland and tunnel flows occur only in a few summer storms (Zhang et al. 1992; Luk 1991). A long term rainfall record shows that the mean event precipitation of runoff-generating storms was 21.5 mm and the mean maximum 30-min rainfall intensity was 19.2 mm/h with a mean duration of 3.35 h (SISWC 1982).

Topographically, the hill slopes in Yangdaogou are characterized by gentle upper slopes of <5°, intermediate to steep middle slopes of up to 25°, and very steep lower slopes of >30°. The upper and middle slopes are mainly comprised of the croplands, while the lower slopes were bush-lands or barren valley slopes.

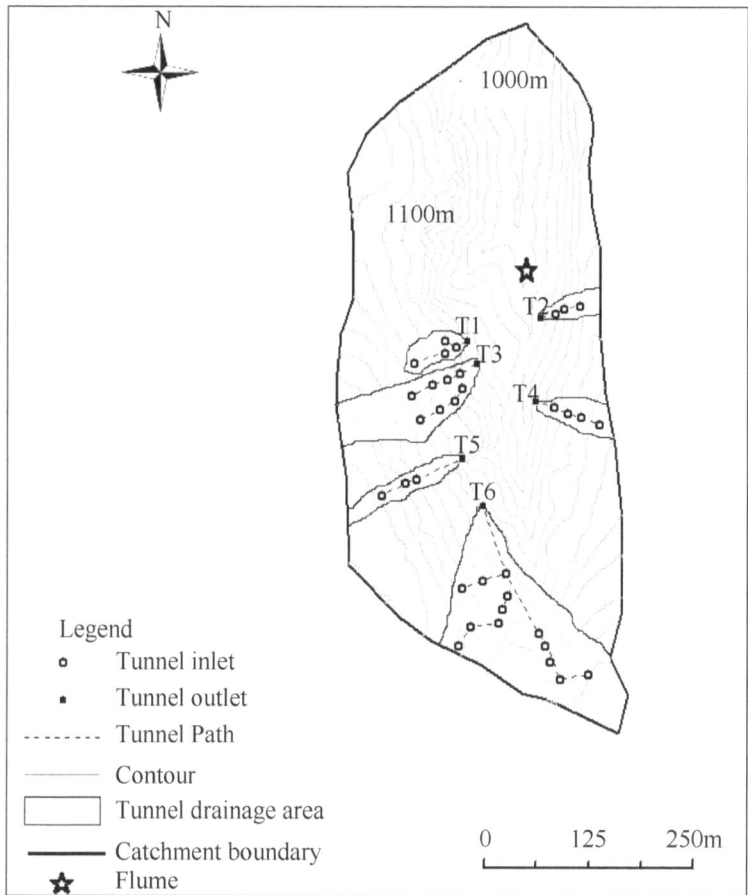

Fig. 13.1 The Yangdaogou experimental catchment, tunnel connectivity and measurement sites. T1–T6 represents the serial numbers of the tunnels

13.3 Field Study Methods

Three parts of a tunnel system were measured in the field: inlet, path and outlet (Fig. 13.2). Most of the tunnel inlets showed a circular and well-like nature, so we chose to measure their diameter and depth to represent their general morphometry. In order to delimit the catchment area of the tunnel systems, tunnel networks were first traced using smoke bombs at the beginning of the monitoring period (Fig. 13.3). During the monitoring period, these tracing experiments were also repeated many times in order to detect their temporal changes. The pressure differential between the tunnel inlet and outlet facilitated this method of evaluation. In order to survey the underground tunnel paths, geophysical methods were applied to measure the electrical resistance of the soils. We infer that tunnel path voids should have extremely

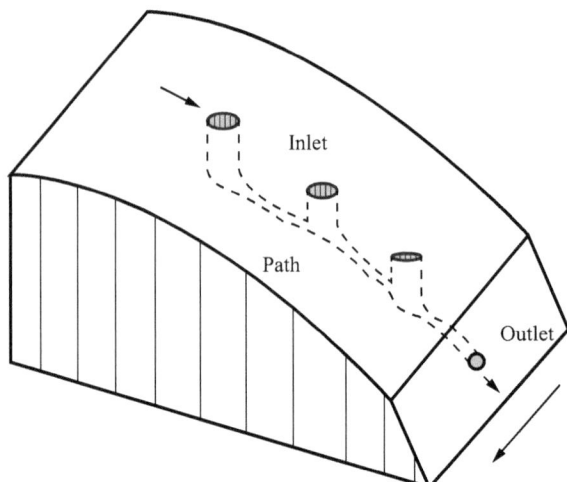

Fig. 13.2 Inlet, path and outlet of a tunnel system

high electrical resistance (Fig. 13.4). The geophysical methods were further validated by the field excavations.

Hydrological and sediment processes of tunnel flows were monitored in 1989 and 1990. To monitor the tunnel flow processes, either flow tanks or simple metal weirs were installed in six tunnel outlets of the Yangdaogou Catchment (Fig. 13.1). The outlets of Tunnel 1, 3, 4 and 6, being located at relatively flat sites, were suitable for the installation of weirs to monitor flow processes. However, the outlets of Tunnel 2 and 5 are located on the cliffs, so that two- and three-order flow divisors were installed there, respectively. Although two tubes with a diameter of 10 cm were used to connect the tunnel outlets with the flow divisors, the tunnel flow mixed with trapped air still readily caused the bricks, mortar seal and tubes to burst a few meters away. After frequent repairs, eventually the monitoring of those two tunnel systems had to be given up, although the start and end time of tunnel flows were still recorded during some events. At the exit of the experimental catchment, a concrete flume was constructed by Hamilton (1990) to check the outflows. Additionally, for the purpose of comparison between overland flow and tunnel flow, five surface plots, with areas from 8 to 21,500 m^2, were established in the Yangdaogou Catchment, but only two of the plots were monitored for processes. Owing to the high sediment concentration in the flows, automation of runoff and sediment monitoring was difficult. Thus, stage readings were manually taken every minute throughout the flow events and sediment samples were taken every three minutes during the first half-hour and every six minutes during the second half-hour and every twelve minutes thereafter. The stage readings were first converted into discharge using the formulae developed and tested by Zeng (1983) and then sediment discharges in the flow were further excluded.

Fig. 13.3   Using smoke bombs to trace tunnel connectivity

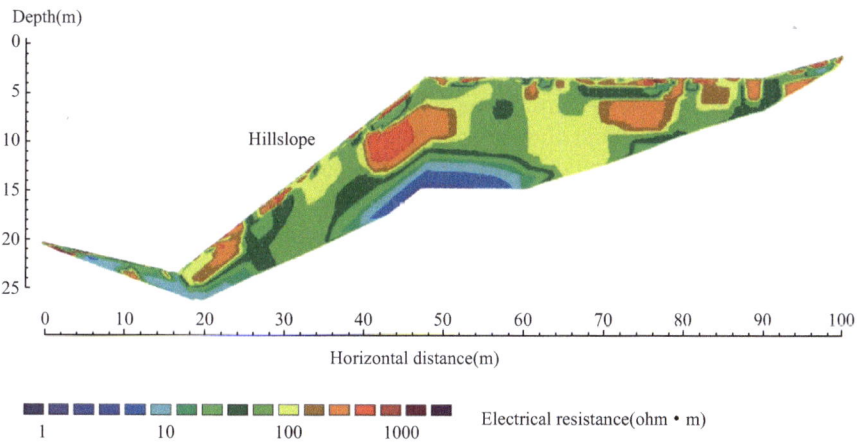

Fig. 13.4   Electrical resistance measured along a soil profile in which a tunnel typically shows high electrical resistance

13.4 Results and Discussion

13.4.1 Hydrological Processes of Tunnel Flows

During 1989 and 1990, tunnel flow occurred in twelve of the fifteen runoff-generation storms. Monitoring the responses to natural storms in arid and semiarid regions was always problematic due to the low frequency of events. In addition, most of the storms occurred at night. This was not an unusual characteristic of storms in these regions, but it did pose greater difficulties for all the assistants to arrive at the monitoring stations at the beginning of the storms. The consequence of all these was that a complete record of each storm at each tunnel outlet was virtually impossible to obtain. In spite of all the logistical difficulties, reasonable data were available from four tunnels (1, 3, 4 and 6). In particular, the largest systems, Tunnels 3 and 6, as well as the catchment outlet had completed records of runoff and sediment data in most of the storms. Thus, it was possible to estimate the contribution of tunnel flows to catchment sediment yield.

(1) Hydrological response of tunnel flows

Tunnel flow start time is here defined as the lag time between the onset of rainfall and the emergence of flow at the tunnel outlet. During 1989 and 1990, tunnel flows occurred in twelve of the fifteen monitored storms. Overall, their response to rainfall was very fast, with the start lag times ranging from 1 to 67 min (Table 13.1). No distinctive difference exists between tunnel flow start time and overland flow start time, which implies that tunnel flow velocity is very fast. Although no measurements were made, we could roughly estimate it. The shortest start time for Tunnel 6 was 5 min. Even if we assume that overland flow was initiated immediately and entered the last tunnel inlet after the rainfall start, the tunnel flow still had an average velocity of 0.26 m/s. It is noted that the tunnel length of this section was actually measured through crawling through by me.

A total of 42 discharge peaks were observed from 35 complete process records at four tunnels (1, 3, 4, 6). Most tunnel hydrographs only had a single peak, which occurred within half an hour after the start of tunnel flow. The number of peaks seems to be determined by rainfall characteristics and is not affected by tunnel complexity. However, exceptions do exist. For example, the storm on July 11, 1990 was characterized by a single rainfall peak (Fig. 13.5). Likewise, one discharge peak occurred in Tunnel 1, 4 and 6. However, two tunnel flow peaks were produced in Tunnel 3. The reason is still unclear. One possible explanation is that this is due to the runoff generation zonation. In light or medium storms, runoff in the tunnel catchment is mainly generated from the cultivated slopes and the steeper valley side slopes and could be quickly directed into the tunnel system. However, in heavy storms such as the above-mentioned one, a large amount of runoff could also be generated from the terrace lands and took a considerably longer time to reach the tunnel inlets than the former owing to the relatively remote locations and gentle slopes. Therefore, it

Table 13.1 Tunnel and catchment flow start time and duration

Date	Rainfall (mm)	Flow start time after rainfall (min)								Flow duration (min)							
		T1	T2	T3	T4	T5	T6	S1	C	T1	T2	T3	T4	T5	T6	S1	C
1989-08-06	24.8	10	16	4	8.7	13	5	M	13	25	29	58	32	31	41	M	61
1989-08-10	13.1 N	N	N	N	N	N	N	M	25	N	N	N	N	N	N	M	25
1989-08-15	28.7	21.4	4.2	18	17.5	M	20	M	15	21	41	37	45	M	40	M	165
1989-08-16	29.1	N	N	N	N	N	N	M	10	N	N	N	N	N	N	M	445
1990-07-06	21	N	N	N	N	N	N	110	M	N	N	N	N	N	N	M	M
1990-07-07	9.5	12	15	13	12	M	23	13	3	11	10	7	6	M	9	17	34
1990-07-11	39.7	15	8	11	14	8	17	12	11	37	9	96	51	M	57	56	96
1990-07-13	28.5	49	29	59	12	29	15	52	27	31	14	M	28	M	45	51	60
1990-07-22	18.2	67	40	M	30	20	10	73	12	51	41	71	55	M	57	56	96
1990-07-26a	33.3	19	1.0	9	11	8	17	6	7	60	46	103	120	50	48	108	105
1990-07-26b	19.8	M	M	M	M	M	18	M	6	M	M	M	M	M	27	M	141
1990-07-30	15.8	M	22	30	25	M	12	15	17	M	25	26	29	M	29	34	60
1990-08-11	20.2	20	17	23	20	M	20	18.3	21	24	21	32	28	M	39	36	53
1990-08-13	35.4	M	29	M	13	M	15	21	M	M	40	M	77	M	52	38	M
1990-08-28	53	13	15	23	17	M	15	M	20	62	42	92	83	M	41	42	63

Notes T = tunnel system; S = surface plot; C = catchment outlet; M = missed events; N = no runoff occurs

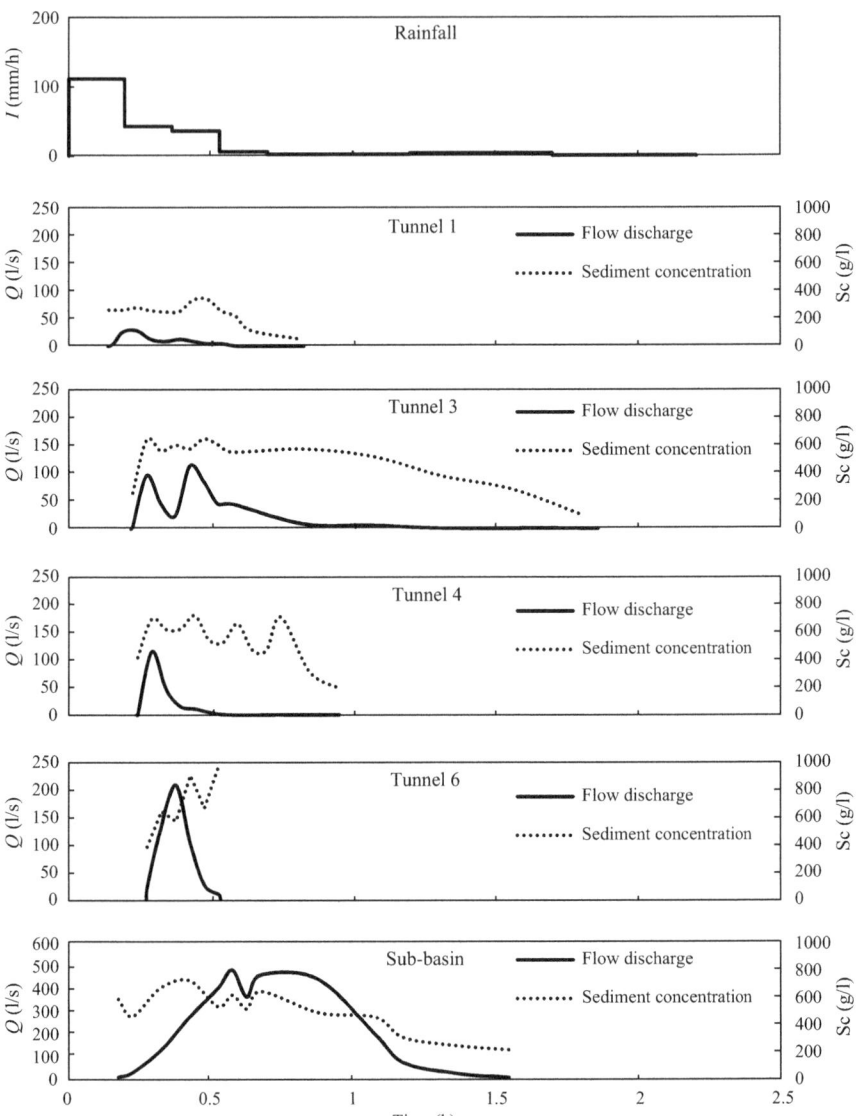

Fig. 13.5 The Yangdaogou experimental catchment, tunnel connectivity and measurement sites

formed a second discharge peak for the same rainfall peak. The uneven peaks in the flume hydrograph implied the slight difference in peak time among different tunnels.

In light and medium storms, tunnel flow duration showed only limited differences among the tunnels, and was more or less comparable to the effective rainfall duration (Table 13.1). However, in the heavy storms, such as on July 11, July 26, and August 28, 1990, Tunnel 3 and 4 had a considerably longer flow duration than the others.

This is because the runoff could be generated from the terrace lands in heavy storms and it took quite a long time to enter the tunnel networks for Tunnel 3 and 4, whereas the tunnel flow duration for the remaining ones was still quite similar to that in light and medium storms, owing either to relatively shorter distances between terrace land and nearest inlets or to higher slope gradients in the tunnel catchments.

(2) Impacts of instability within tunnel systems on tunnel flow hydrology

Deep-seated tunnel systems in this area are characterized by great instability. Collapses within tunnel systems are very common. Small-scale collapses may only cause oscillation of sediment concentrations in the tunnel flow and no effects on tunnel flow hydrology. However, large collapses could exert profound impacts on tunnel flow hydrology. If the collapses are extremely large or associated with surface depression or sediment deposition, tunnel systems could be totally blocked. Rapid tunnel flow could, in turn, reopen the blocked tunnel systems. Such a temporal shift of tunnel systems can be detected using smoke bombs before and after storm events. In the 1989 rainy season, the outlet of Tunnel 1 was connected to two series of inlets, but the major branch was blocked in the 1990 rainy season. This led to a great disparity in tunnel flow discharge between 1989 and 1990. In Tunnel 3, the southern branch, consisting of four tunnel inlets, was blocked throughout the 1989 and 1990 rainy seasons, but it was reopened in 1992. The middle branch was also blocked during the storm of August 11, 1990 and reopened by the storm of August 28, 1990, which caused the tunnel flow discharge of the former storm to be disproportionally low. In Tunnel 4, three connecting tunnel inlets, about 30 m south of the tunnel outlet, were abruptly joined into Tunnel 4 during the storm of August 13, 1990. Those newly joined tunnel inlets subsequently added a large amount of runoff to the tunnel system generated from slope land and terrace land. The most significant event which the authors observed during the two consecutive rainy seasons was the abrupt initiation of one tunnel inlet on August 13, 1990. The inlet, with a diameter of 1.5 m and depth of 1.9 m, was developed in the middle of a road, located on the upper drainage boundary of Tunnel 6. Smoke bomb tests indicated that it was connected to an inlet of Tunnel 6 about 40 m away. The runoff from the village and the neighboring sub-basin, which used to flow into another basin via the excavated road, was redirected into Tunnel 6 through the inlet and conduit of the newly developed tunnel. This led to the discharge of Tunnel 6 being unusually high during the storm of August 13, 1990. After that storm, the inlet was filled in by the villagers, since it hindered traffic. As a result, discharge at Tunnel 6 returned to normal in the subsequent storm of August 28, 1990.

Totally blocked tunnel branches can be readily detected with smoke bombs and thereby their impacts on tunnel flow hydrology can be explicitly evaluated. However, in most cases, the tunnel systems may not be totally blocked but partially dammed or blocked first and reopened later during the same storm. In these situations, smoke bombs are useless and it is extremely dangerous to investigate by crawling into the tunnel systems after storms. Thus normally no direct evidence is available. However, in July of 1989, the authors did manage to pass through the last section of Tunnel 6, a 76 m long tunnel conduit, and found a bridge-like constriction inside. Apparently, it used to be a dam caused by a collapse and sometime later tunnel flow penetrated the dam and formed the opening under the 'bridge'. Here, in contrast to the intended objective of this section, we use the monitored tunnel flow hydrologic processes to identify the possible partial damming during the event. Owing to the lack of direct evidence, the results presented here must be considered tentative and to be examined further in the future. In the storm of July 26, 1990, tunnel flow processes in all tunnels except Tunnel 4 were characterized by an early discharge peak, which was caused by an immediate intensity peak after rainfall onset. However, discharge at Tunnel 4 was very low in the first hour and peak discharge did not occur until 77 min after rainfall onset. After the peak, the discharge sharply dropped to a very low level and lasted for another 40 min or so. The total discharge appeared to be normal and the absence of the first peak probably resulted from partial damming, which was flushed away later by accumulated water inside the tunnel. In contrast, in Tunnel 6, after the first discharge peak, tunnel flow simply stopped. It was unlikely that no runoff had been generated by the second rainfall peak from the tunnel catchment, the largest one in the basin. The total discharge from Tunnel 6 during this storm was also quite low. Twelve hours later, another storm occurred, with a rainfall of 19.8 mm. Flow in Tunnel 6 started 18 min after rainfall onset and the discharge was so high that the trapezoidal weir shifted. Accordingly, no water discharge data were collected, although sediment samples were still taken throughout the event. Actually, the mean rainfall intensity for the first 20 min (0.09 mm/min) was very low in this storm. Though the antecedent soil moisture was very high, it is unlikely to have produced such a high flow if no water had been trapped by damming during the previous storm. Two rainfall peaks produced four discharge peaks in Tunnel 3, may have been caused by runoff generation zonation as well. To further evaluate the impact of instability on tunnel flow discharge, the authors compared it with overland flow discharge. Owing to the limited number of events that were monitored on the surface plots during our monitoring periods, the authors have used the data collected by the Shanxi Institute of Soil and Water Conservation during the period of 1963–1968 from the Yangdaogou sub-basin (SISWC 1982). Three surface plots with areas of 400,1855 and 4167 m^2 were selected in the sub-basin comprising upper slope, lower slope, and combined slope, respectively. It was found that a good correlation exists between runoff discharge and rainfall with an intensity of more than 0.2 mm/min for all three surface plots. For tunnel systems, such a good correlation could only be found for Tunnel 3. However, if we disregard those events affected by tunnel instability, identified above, correlation coefficients are improved for all tunnels, especially for Tunnel 1 and 4. The still poor correlation for Tunnel 6 may be ascribed to unidentified tunnel instabilities within this large and complex system.

13.4.2 Sediment Processes of Tunnel Flows

(1) Within storm variations

To examine the relationship between sediment and runoff discharge, the correlation coefficients between measured flow discharge rates and sediment concentrations for all recorded events at Tunnel 1, 3, 4 and 6, as well as at catchment outlet are determined. Sediment concentration shows a good correlation with discharge only in 20% of all recorded tunnel flow events, but in 60% of catchment outflow events. Among those tunnel flow events with a significant correlation between runoff discharge and sediment concentration, five occurred at Tunnel 3, one each in Tunnel 1 and Tunnel 4, and none in Tunnel 6. It is noted that the best correlation between discharge and sediment concentration does not exist in small tunnels such as Tunnel 1 or Tunnel 4 but in Tunnel 3, the second-largest tunnel in the catchment. This implies that the correlation between runoff discharge and sediment concentration is not directly related to the size of tunnel systems but the tunnel stability.

To further examine the erratic relationship between discharge and sediment concentration, timing of peak flow discharge is compared with that of peak sediment concentration for each tunnel flow event. Overall, 65% of events show the first peak sediment concentration precedes the first peak runoff. A wide variation also exists between tunnels, with 63% at Tunnel 1, 22% at Tunnel 3, 88% at tunnel 4, and 88% at Tunnel 6, respectively. The early peaking of sediment concentration suggests that there are significant flushing effects in these tunnel systems. The flushing could relate to the sediment deposition in the previous storm or collapse during the inter-storm period. For early storms of the season, it may relate to deposition from a range of processes (mass wasting, aeolian deposition, etc.) during the previous fall and winter. This view is supported by observations of a substantial thickness of fluvial deposits and wind-blown dusts in the sloping section of Tunnel 6 and the roof collapse materials of Tunnel 1 prior to the occurrence of the first major storm of 1990.

Tunnel flow hydrographs are characterized by a rapid recession limb, but sediment concentration remains very high in many events. This is reflected by a counter-clockwise loop in the later part of the event in hysteresis graphs (Fig. 13.6). Of a total of 35 tunnel flow events, 19 or 56% have a counter-clockwise loop in the latter part of the event. This pattern was found to be associated with the occurrence of hyperconcentrated flow which maintained high sediment concentration during flow recession (Hamilton 1990; Qian and Wan 1983; Wang et al. 1982).

Comparison of data from tunnel outflow and catchment outflow indicates that peak sediment concentration in tunnel-flows is not distinctively higher than that in the catchment outflow. This is different from the findings by Bryan and Harvey (1985) in the Alberta Badlands where peak sediment concentration of pipe flow is substantially above peak channel concentrations. It is noted that the measured peak sediment concentrations in this area were one magnitude order higher than those measured in the Alberta Badlands area. However, except for Tunnel 1, peak sediment

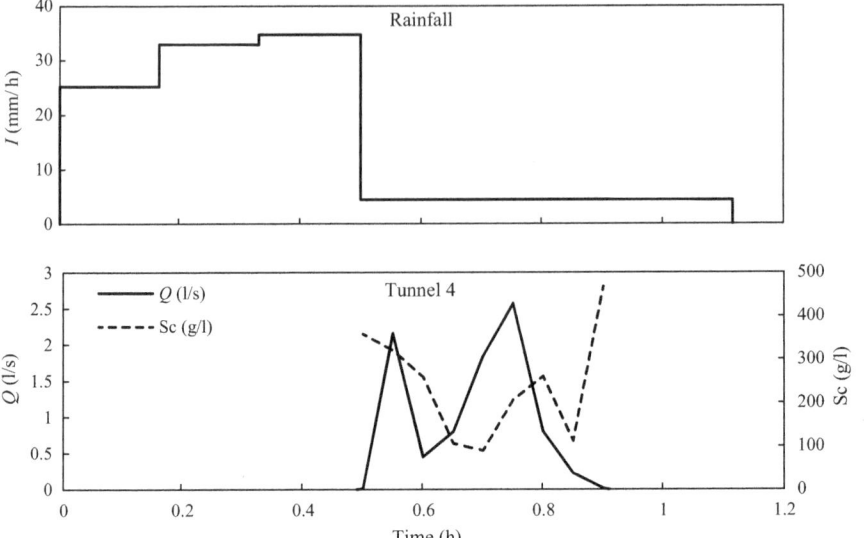

Fig. 13.6 Water discharge, sediment concentration tunnel flow in Tunnel 4 in the storm of July 20, 1990

concentrations of the tunnel flows are considerably higher than those measured from the un-tunneling hill slopes.

(2)   Between storm variations

To evaluate sediment variations between storms, discharge-weighted mean sediment concentration (MSC) was determined for each storm. The absolute range of MSC obtained is from 34 to 671 g/l. MSC was then plotted against total discharge per storm for each tunnel outlet (Fig. 13.7). None of the tunnels shows significant relations between flow discharges and discharge-weighted mean sediment concentration (MSC).

Study of time series data of MSC allows identification of possible seasonal patterns. Here, variations of MSC were examined as a time series for each tunnel outlet (Fig. 13.7). It can be seen that the temporal change of MSC is very different between tunnel systems and it is very difficult to identify any trends which can be generalized. As we missed monitoring the storms of the early rainy season in 1989, variations of MSC in 1990 were examined for seasonal patterns for Tunnel 1, 3, 4 and 6. In Tunnel 1, there is a general decline in MSC throughout 1990. The MSC of the July 7, 1990 event, 540 kg/m^3, is well above those of the rest of the storms during the season. The highest MSC in the first storm was due to the "preparation" of materials by mass wasting prior to the rainy season. A section of tunnel roof was collapsed in the winter of 1989. Most of the debris materials were flushed away in the storm of July 7. However, the difference in MSC between the first and the rest of the storms in large tunnel systems seems not to be as great as in Tunnel 1. The more pronounced

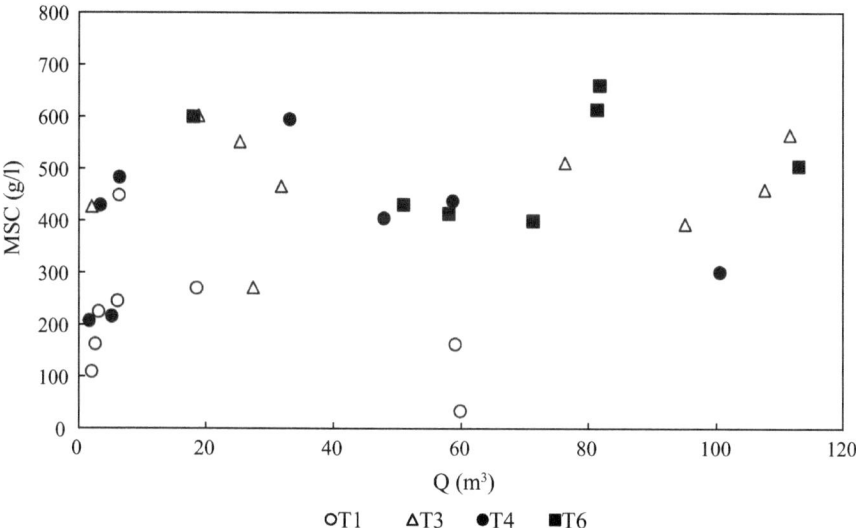

Fig. 13.7 Correlation between MSC and flow discharge per storm for tunnel flows

"flush effect" in the small tunnel systems may result from lower dilution effects due to small drainage areas. Over the field season, Tunnel 3 shows limited variations in MSC and the temporal changes of MSC in Tunnel 4 and 6 are very irregular.

One of the identifiable factors to affect temporal changes of MSC is the variable source of tunnel runoff and sediment. While this is also significant in small basin studies, the situation regarding tunnels is more complicated because of rapid temporal changes in the sources. In Tunnel 1, the contributing area was considerably smaller in the 1990 storms than in the 1989 storms due to blockage of a major inlet. Accordingly, much lower runoff discharge was recorded in all the storms in 1990 but the MSC increased dramatically due to lower dilution effects. In Tunnel 4, the highest MSC occurred in the August 28, 1990 event, which was ascribed to the abrupt addition of a new section of tunnel in the late period of the August 13 storm event. In Tunnel 6, on August 13, a new inlet developed in the middle of a road, located on the upper drainage boundary, which diverted runoff from village and roads into Tunnel 6. This led to the discharge of Tunnel 6 being unusually high during the August 13 storm but to decrease in MSC due to dilution effects. Overall, tunnel instability such as blockage, roof collapse and opening of new inlets seems to be responsible for the erratic relationship between MSC and flow discharges.

Table 13.2 Flow and sediment contributions to the catchment by tunnels during the monitoring period

Date	Rainfall (mm)	Tunnel flow discharge (m^3)	Tunnel flow contribution to catchment (%)	Tunnel sediment discharge (kg)	Sediment contribution of tunnels to catchment (%)
1989/8/6	24.8	224.0	50.7	114,126	47.0
1989/8/10	13.1	0	0	0	0
1989/8/15	28.7	221.8	58.9	93,685	80.1
1989/8/16	29.1	0	0	0	0
1990/7/6	21.0	0	0	0	0
1990/7/7	9.5	19.3	7.2	7400	13.6
1990/7/11	39.7	218.2	20.2	236,577	55.7
1990/7/13	28.5	153.6	49.8	88,482	66.5
1990/7/22	18.2	104.5	39.1	30,106	49.1
1990/7/26	33.3	164.6	40.9	81,571	62.6
1990/7/30	15.8	85.5	77.1	53,202	76.9
1990/8/11	20.2	23.9	25.0	13,742	62.4
1990/8/13	35.4	486.6	–	135,689	–
1990/8/28	53.0	219.8	48.8	87,504	56.2
Total	334.9	1921.8	42.9	942,084	57.0

13.4.3   Tunnel Sediment Contributions to Catchment Yield

During the monitoring period, there were fifteen storms producing channel flows at the catchment outlet. Sediment yields were only measured from Tunnel 1, 3, 4 and 6, but they account for about 90% of the drainage area of all tunnel systems in the catchment. The results are listed in Table 13.2.

Overall, about 47% of catchment flow is contributed from tunnel systems. The contributions to the catchment outflow are often significantly high in medium storms such as on August 15, 1989 and on July 22 and 30, 1990. In very heavy rainstorms, such as July 11, 26 and 28, 1990, tunnel flows contributions to the catchment outflow fall off, and are more or less comparable to the proportion of tunnel drainage area to the catchment area.

About 57% of total catchment sediment production is delivered by the tunnel systems. The sediment contributions widely vary between storms, ranging from 0 to 80%. Though the tunnel instability makes the relationships between runoff and sediment yield erratic at most tunnels at both within- and between-storm levels, the overall sediment contributions by tunnel systems to catchment yield show a general pattern at the catchment level. In light storms with low intensities such as on August

10, 16, 1989 and July 6, 1990, no tunnel flow was generated, though catchment out-flow was observed. In medium storms, sediment contributions to the catchment by the tunnel systems are often high such as on August 15, 1989 and July 30, 1990. However, in heavy storms such as July 11, 26 and August 28, 1990, sediment contri-butions of tunnel flows fall off and are slightly higher than the proportion of tunnel catchment areas to the catchment. In general, the pattern of sediment contribution is quite similar to that of water discharge contribution (Zhu 1997).

13.5 Conclusions

A monitoring of tunnel flows was carried out in the Yangdaogou Catchment over the 1989 and 1990. Although tunnel flows were derived solely from the surface catch-ment via tunnel inlets, they did not simply mirror overland flow processes due to instability within tunnel systems. Moreover, the peak sediment concentrations in the tunnel flows were not distinctively higher than the peak channel flow concentrations but considerably higher than those measured from the un-tunnelled hill slopes. In general, no significant correlation between the flow discharges and sediment yields could be found for the tunnel flows. Such an erratic relationship may be ascribed to the influence of the variable sediment source area, the occurrences of collapses within tunnel systems, and the initiation of new inlets. The temporal changes of sed-iment yield on an annual basis were very irregular and no seasonal trends might be generalized for all tunnels. Based on the field monitoring data, about 43% of catch-ment runoff and 57% of catchment sediment were delivered by the tunnel systems. These results clearly show that the tunnel system played a major hydrologic and geomorphic role in the hilly loess region of the Western North China.

References

Boucher S C, Powell J M. 1994. Gullying and tunnel erosion in Victoria. Australian Geographical Studies, 32(1): 17–26.

Bryan R B, Harvey L E. 1985. Observations on the geomorphologic significance of tunnel erosion in a semi-arid ephemeral drainage system. Geografiska Annaler: Series A, Physical Geography, 67(3–4): 257–272.

Bryan R B, Jones J A A. 1997. The significance of soil piping processes: inventory and prospect. Geomorphology, 20(3–4): 209–218.

Carey S K, Woo M K. 2000. The role of soil pipes as a slope runoff mechanism, Subarctic Yukon, Canada. Journal of Hydrology, 233(1–4): 206–222.

Chen Y Z, Luk S H. 1989. Sediment sources and recent changes in the sediment load of the Yellow River, China. Proceedings of 5th International Soil Conservation Conference, Bangkok: 313–323.

Crouch R J, McGarity J W, Storrier R R. 1986. Tunnel formation processes in the Riverina area of NSW, Australia. Earth Surface Processes and Landforms, 11(2): 157–168.

Dardis G F, Beckedahl H R. 1988. Drainage evolution in an ephemeral soil-pipe gully system, Transkei, Southern Africa// Dardis G F, Moon B P. Geomorphological Studies in Southern Africa: Rotterdam, Balkema: 247–265.

Drew D P. 1982. Piping in the big muddy badlands, southern Saskatchewan, Canada// Bryan R, Yair A. Badland Geomorphology and Piping, Norwich: GeoBook: 293–304.

Faulkner H, Alexander R, Teeuw R, et al. 2004. Variations in soil dispersivity across a gully head displaying shallow sub-surface pipes, and the role of shallow pipes in rill initiation. Earth Surface Processes and Landforms, 29(9): 1143–1160.

Frankl A, Deckers J, Moulaert L, et al. 2016. Integrated solutions for combating gully erosion in areas prone to soil piping: innovations from the drylands of Northern Ethiopia. Land Degradation and Development, 27(8): 1797–1804.

Fuller M L. 1922. Some unusual erosion features in the loess of China. Geographical Review, 12(4): 570–584.

Gilman K, Newson M D. 1980. Soil pipes and pipeflows—a hydrological study in upland wales. Norwich: Geobooks: 114.

Hamilton H. 1990. Field examination of soil erosion and losses of nitrogen and phosphorus from the agricultural watershed in the Loess Plateau, People's Republic of China. Canada: University of Toronto: 205.

Harvey A. 1982. The role of piping in the development of badlands and gully systems in south-east Spain// Bryan R, Yair A. Badland Geomorphology and Piping. Norwich: GeoBook: 317–335.

Heede B H. 1971. Characteristics and processes of soil piping in gullies. U.S. Department of Agriculture, Forest Service Research Paper, RM-68: 15.

Holden J. 2005. Controls of soil pipe frequency in upland blanket peat. Journal of Geophysical Research: Earth Surface, 110(F1): F01002.

Jones J A A. 1981. The nature of soil piping: a review of research// British Geomorphological Research Group Research Monograph No.3. Norwich: GeoBooks: 301.

Jones J A A. 1987. The effects of soil piping on contributing areas and erosion patterns. Earth Surface Processes and Landforms, 12(3): 229–248.

Jones J A A, Crane F G. 1984. Pipeflow and pipe erosion in the Maesnant experimental catchment// Burt T P, Walling D E. Catchment Experiments in Fluvial Geomorphology. Norwich: GeoBooks: 55–72.

Jones J A A, Bryan R B. 1997. Piping erosion. Special Issue in Geomorphology: 319.

Liu D S. 1964. Loess deposits in the middle reaches of Yellow River. Beijing: Science Press: 234 (in Chinese).

Luk S H. 1991. Soil erosion and land management in the Loess Plateau Region, North China. Chinese Geography and Environment, 3: 3–28.

McCaig M. 1983. Contributions to storm quickflow in a small headwater catchment - the role of natural pipes and soil macropores. Earth Surface Processes and Landforms, 8(3): 239–252.

Onda Y. 1994. Seepage erosion and its implication to the formation of amphitheatre valley heads: a case study at Obara, Japan. Earth Surface Processes and Landforms, 19(7): 627–640.

Parker G G, Jenne E A. 1967. Structural failure of western highways caused by piping. Highway Research Record, 203.

Parker G G, Higgins C G, Wood W W. 1990. Piping and pseudokarst in drylands. Geological Society of America Special Paper, 252: 77–110.

Piccarreta M, Faulkner H, Bentivenga M, et al. 2006. The influence of physico-chemical material properties on erosion processes in the badlands of Basilicata, Southern Italy. Geomorphology, 81(3–4): 235–251.

Qian N, Wan Z H. 1983. Sedimentology. Beijing: Science Press: 656 (in Chinese).

Seppälä M. 1997. Piping causing thermokarst in permafrost, Ungava Peninsula, Quebec, Canada. Geomorphology, 20(3–4): 313–319.

SISWC (Shanxi Institute of Soil and Water Conservation). 1982. Rainfall, runoff and sediment data (1955–1981), unpublished (in Chinese).

Swanson M L, Kondolf G M, Boison P J. 1989. An example of rapid gully initiation and extension by subsurface erosion: coastal San Mateo County, California. Geomorphology, 2(4): 393–403.

Thorp J. 1936. Geography of the soils of China// The National Geological Survey of China. Nanking.

Uchida T, Kosugi K, Mizuyama T. 1999. Runoff characteristics of pipeflow and effects of pipeflow on rainfall-runoff phenomena in a mountainous watershed. Journal of Hydrology, 222(1–4): 18–36.

Verachtert E, Van Den Eeckhaut M, Poesen J, et al. 2010. Factors controlling the spatial distribution of soil piping erosion on loess-derived soils: a case study from central Belgium. Geomorphology, 118(3–4): 339–348.

Walsh R P D, Howells K A. 1988. Soil pipes and their role in runoff generation and chemical denudation in a humid tropical catchment in Dominica. Earth Surface Processes and Landforms, 13(1): 9–17.

Wang B K. 1989. Factors causing tunnel erosion. Acta of Soil and Water Conservation, 3(3): 84–90 (in Chinese).

Wang X K, Qian N, Hu W D. 1982. The formation and process of confluence of the flow with hyper-concentration in the Gullied-Hilly Loess areas of the Yellow River Basin. Journal of Hydraulic Engineering, (7): 26–35 (in Chinese).

Wilson C M, Smart P L. 1984. Pipes and pipe flow processes in an upland catchment Wales. Catena, 11(2–3): 145–158.

Wilson G V, Rigby J R, Dabney S M. 2015. Soil pipe collapses in a loess pasture of Goodwin Creek watershed, Mississippi: role of soil properties and past land use. Earth Surface Processes and Landforms, 40(11): 1448–1463.

Yair A, Lavee H, Bryan R B, et al. 1980. Runoff and erosion processes and rates in the Zin Valley badlands, Northern Negev, Israel. Earth Surface Processes and Landforms, 5(3): 205–225.

Zeng B Q. 1983. Methods for soil and water losses monitoring in the small catchment. Soil and Water Conservation of Shanxi, (3): 61–68 (in Chinese).

Zhang Z G, Zheng B Q, Jia Z J. 1992. Precipitation characteristics in Wangjiagou Basin, Western Shanxi// Wang F T. Soil Erosion Management and Application of Geographical Information System on the Loess Plateau Region, Western Shanxi. Beijing: Science Press: 1–12 (in Chinese).

Zhu X M. 1958. Tunnel erosion in the loess region. Construction of the Yellow River, (3): 43–44 (in Chinese).

Zhu T X. 1997. Deep-seated, complex tunnel systems—a hydrological study in a semi-arid catchment, Loess Plateau of China. Geomorphology, 20(3): 255–267.

Zhu T X. 2003. Tunnel development over a 12-year period in a semi-arid catchment of the Loess Plateau, China. Earth Surface Processes and Landforms, 28(5): 507–525.

Zhu T X. 2006. Initiation and development processes of tunnel systems in the hilly loess region of northern China. International Journal of Sediment Research, 21(3): 171–179.

Zhu T X. 2012. Gully and tunnel erosion in the hilly Loess Plateau region, China. Geomorphology, 153–154(1): 144–155.

Zhu T X, Cai Q G, Luk S H. 2000. Contribution of tunnel erosion to sediment yield in a semi-arid catchment. International Journal of Sediment Research, 15(4): 440–444.

Zhu T X, Luk S H, Cai Q G. 2002. Tunnel erosion and sediment production in the hilly loess region, North China. Journal of Hydrology, 257(1–4): 78–90.

Additional Material

Additional table   Patents invented by the authors, each of which the first inventor is Xiangzhou Xu. All patent technologies listed in the following table have been applied in the soil conservation experiments, and the anticipated results have been obtained.

X. Xu et al., *Experimental Erosion*,
https://doi.org/10.1007/978-981-15-3801-8

No.	State Intellectual Property Office, China		European Patent Office (https://www.epo.org/index.html)	
	Inventor(s). Title (Grant No.)	Bibliographic data	Webpage and abstract	
1	Xu X Z, Guo W Z, Zhang H W, Gao L, Zhao X Y. Topography observation device based on mechanical fine tuning dual hosts	CN108534758 (A)—2018-04-08	The invention belongs to the technical field of water and soil conservation study devices, and relates to a dynamic observation device used for a slope topography evolution process. The two hosts emit parallel, equal-height and overlapping laser planes from different orientations, and the laser planes are projected onto the slope topography; an image collecting device shoots the slope topography which laser rays are projected onto at an angle perpendicular to the layer planes to form a three-dimensional stereogram, and then parameters such as the slope volume and the slope gradient are calculated. According to the device, by rotating rotary knobs of mechanical fine tuning devices in the topographic meter hosts, equal-interval parallel accurate adjustment of laser ray planes is achieved; according to the device, by observing a fixing bolt on a support and a worm handle rotary knob in a view screen box, moving and rotation of a camera used for image collection are controlled. The invention provides a novel topographic meter through which the spacing and the angle of a linear laser can be conveniently calibrated, and all the topographies including local deep grooves on an eroded slope can be accurately observed	
2	Xu X Z, Zhang C, Liu Y K, Wang R Q, Song G D. A 3D sampling device for the pollutants transporting in a river (ZL201511024920.8)	CN105403435 (A)—2016-03-16	The invention provides a three-dimensional sampling device for simulating transport of river contaminations. The three-dimensional sampling device is composed of a bank control point, a catamaran sampling platform and a multi-point sampler. The device is characterized in that the catamaran sampling platform moves in a river under the control of the bank control point; and when the catamaran sampling platform reaches a sampling position, the multi-point sampler installed on a sampling device takes water samples at different water depths and different distances from an initial point from a to-be-measured section. The three-dimensional sampling device provided by the invention can meet requirements for safety, easy operation, small disturbance to a flow field of river water and the like at the same time, has high sampling efficiency and accurate sampling position and guarantees accuracy and representativeness of the water samples	

(continued)

(continued)

No.	State Intellectual Property Office, China		European Patent Office (https://www.epo.org/index.html)	
	Inventor(s). Title (Grant No.)	Bibliographic data	Webpage and abstract	
3	Xu X Z, Zhao R. Device and method for field experiments of the gravity erosion on the gully wall (ZL201310422836.6)	CN103487567 (A)—2014-01-01	The invention discloses a device and method for building a trench slope gravity erosion process field test. The device comprises an outdoor test room, trench slope geography, a rainfall simulation device and a gravity erosion observation device. The outdoor test room comprises an outdoor test room body capable of being assembled and a movable rainproof observation room body placed inside the outdoor test room body. The trench slope geography is a generalization slope cut in a field according to typical geomorphic features of a studied area. The generalization slope is separated from the surrounding through steel sheet pile side walls. The rainfall simulation device comprises components including a three-barrel pressure-stabilizing water feeder, a movable assembling multi-micro-spraying-nozzle combined type rainfall device, a water collecting groove and the like.; The gravity erosion observation device comprises a geography instrument main machine, a movable assembling slumping face observation support and a sight line calibrator. According to the device and method for building the trench slope gravity erosion process field test, a good observation environment is created for observation of a geography instrument through the test room, the main building components are all assembling movable facilities, the simulation and observation devices can not be disturbed, and observation accuracy is guaranteed	

(continued)

(continued)

No.	State Intellectual Property Office, China		European Patent Office (https://www.epo.org/index.html)	
	Inventor(s). Title (Grant No.)	Bibliographic data	Webpage and abstract	
4	Xu X Z, Zhu X B, Liu L, Ma L, Wang P P. A rainwater collection device based on concrete sand-based porous bricks (ZL201310237771.8)	CN103352411 (A)—2013-10-16	The invention relates to a rainwater penetration collection device based on concrete-sand-based water-permeable bricks. The invention belongs to the field of construction design. The invention is characterized in that: pedestrian walkways and squares are paved by using concrete-sand-based water-permeable bricks; rainwater enters the concrete-sand-based pavement water-permeable bricks and cover-plate water-permeable bricks, and is filtered by a gravel bedding layer and a crushed stone bedding layer; rainwater enters a rainwater collection ditch through the water-permeable bricks of a ditch wall of a rainwater-collection thin ditch side wall; the rainwater is then gathered to a water storage tank on the square and is used as a landscape or used for irrigation. The device is characterized in that screened and processed desert sand is used as a main aggregate; high-grade cement is adopted as an adhesive; and proper additives are added. Frustum-shaped water-permeation holes are provided on brick bodies, and anti-slip stripes are arranged on the brick body. The device provided by the invention has the advantages that: desert sand is sufficiently utilized; the strength is high, and cost is low; the water permeability is good, and the water-permeation holes are prevented from blockage; the device is safe and comfortable; a good rainwater collection plane which is the square is sufficiently utilized; and rainwater processing is simple and feasible	

(continued)

(continued)

No.	State Intellectual Property Office, China		European Patent Office (https://www.epo.org/index.html)
	Inventor(s). Title (Grant No.)	Bibliographic data	Webpage and abstract
5	Xu X Z, Guo W Z, Ma L, Yan Q. A method to observe the amount of gravity erosion in the field experiment (ZL201310422447.3)	CN103487566 (A)—2014-01-01	The invention discloses an observational method used for a trench slope gravitational erosion field test. The observational method is characterized in that a DV sight line is perpendicular to structured light rays on an abrupt slope through a geography instrument, and according to video screenshots in the test process, the gravity erosion amount and the waterpower erosion amount, namely the total erosion amount, the gravity erosion amount and the waterpower erosion amount of individual rainfall, in an individual rainfall event, the gravity erosion amount occurring after rainfall is completed, and slope direction distribution and gradient distribution of the abrupt slope at any time can be respectively calculated.; The observational method is further characterized in that the definition of laser rays projected on the abrupt slope is strengthened by adjusting indoor light rays, building a rainproof observation room and installing an anti-fog filter, the DV sight line is perpendicular to a laser plane emitted by a geography instrument main machine by adjusting the sight line angles of a DV, and testing and adjustment are carried out according to a sight line calibrator. According to the observational method, the water power erosion amount can be separated from the gravity erosion amount, and the gravity erosion amount in the rainfall process and the gravity erosion amount after rainfall can be quantitatively observed and calculated

(continued)

(continued)

No.	State Intellectual Property Office, China	European Patent Office (https://www.epo.org/index.html)	
	Inventor(s). Title (Grant No.)	Bibliographic data	Webpage and abstract
6	Xu X Z, Shu L M, Yang X T, Xu F L. A energy-saving device to be quantitatively observed (ZL201110002354.6)	CN102102378 (A)—2011-06-22	The invention provides an energy-saving rainwater harvesting device capable of quantitative observing, belonging to the technical field of water treatment. The device is suitable for washing cars in the parking lots. The device mainly comprises an initial rainwater discarding device, a physical sedimentation tank and a rainwater harvesting bucket, wherein the initial rainwater discarding device comprises a buffer slot, a COD (chemical oxygen demand) sensor, a CPU, a steering engine and a butterfly valve; the initial rainwater discarding quantity is controlled by a COD concentration detection method; the physical sedimentation tank comprises a buffer plate, a detachable reverse seepage slot, inclined plates, concave slots, drain outlets and the like; the mid-late rainwater is treated by adopting the principle of reverse seepage; and the rainwater harvesting bucket mainly comprises an overflow vent, a water quantity observation instrument, a water receiving hole, a drain outlet, a water pump, an antifreezing film, a solar absorption film and the like. The device has the following beneficial effects: the device can accurately discard, purify and recycle the rainwater according to different degrees of pollution of the rainwater, can quantitatively observe the water quantity and is accurate, flexible, energy-saving and environment-friendly; and the device simultaneously solves the problem of recycle of the rainwater in the districts, architectural complex and regions and can cut down on the pollution load of the urban rainwater and improve the utilization ratio of the rainwater resources

(continued)

(continued)

No.	State Intellectual Property, Office, China		European Patent Office (https://www.epo.org/index.html)	
	Inventor(s). Title (Grant No.)		Bibliographic data	Webpage and abstract
7	Xu X Z, Xu F L, Liu Y K, Wang S F. A 3D observation method for the scouring terrain (ZL201010502055.4)		CN102032902 (A)—2011-04-27	The invention discloses a three-dimensional observation method of eroded landform, belonging to the field of topographic survey. The method is characterized by comprising the steps that: a full-plane laser and a laser positioning spot are emitted to an eroded landform to be observed by a full-plane laser projector and a topography image calibration device; contours are formed on the surface of the landform by the full-plane laser, the laser positioning spot is used as a control spot for dimensional calibration and deformational correction of topography images; the full-plane laser projector is translated or lifted by certain steps, the landform images with different initial altitudes are acquired by an image acquisition device and then superposed and fused via computer software, so as to realize the dense acquisition of landform surface spots; and three-dimensional coordinates of the landform surface spots are acquired in accordance with geometrical information of laser lines in the images to further obtain a micro-space digital landform contour map, thus noncontact three-dimensional measurement for the eroded landform and computational analysis for the three-dimensional model thereof are achieved. The entire system has high observation efficiency and accuracy, and is suitable for the three-dimensional observation of landform in river engineering model test

(continued)

(continued)

No.	State Intellectual Property Office, China	European Patent Office (https://www.epo.org/index.html)	
	Inventor(s). Title (Grant No.)	Bibliographic data	Webpage and abstract
8	Xu X Z, Xu F L, Wang J, Liu Y K, Wang S F. A 3D observation device for the scouring terrain (ZL201010502051.6)	CN101975570 (A)—2011-02-16	The invention relates to a three-dimensional observation device for scouring terrain, belonging to the field of terrain measurement and comprising a full-plane laser projector, an image collection device, a terrain image calibration device, a host lifting device, a fixed support structure and a power supply system, wherein the full-plane laser projector comprises a slot-shaped line type laser die set, a die set micro-adjusting bracket, a baffle board, a cover board, a laser distance meter and a universal joint; the terrain image calibration device comprises a spot type laser die set and a fixed supporting board; the image collection device comprises a computer, a digital camera, a camera gravity positioning box and a universal joint; the respective horizontal angle of the full-plane laser projector and the digital camera is automatically adjusted through the universal joint under the self-weight action so that the full-plane laser is in parallel to a horizontal plane; the principal optical axis of the camera is perpendicular to the horizontal plane; and the host lifting device controls the lifting of the full-plane laser projector in a step pitch setting mode. The integral set of the system has higher observation efficiency and precision, and is suitable for the research on the aspect of terrain three-dimensional analysis in a river model test

(continued)

(continued)

No.	State Intellectual Property Office, China		European Patent Office (https://www.epo.org/index.html)
	Inventor(s). Title (Grant No.)	Bibliographic data	Webpage and abstract
9	Xu X Z, Xu F L, Zhao C, Xu Y, Guo X Y. A measurement method to model the 3D landform (ZL201010144689.7)	CN101865690 (A)—2010-10-20	The invention relates to a method capable of observing the three-dimensional ditch and slop micro topographical feature in real time, which belongs to the field of topographic form measurement. The invention is characterized in that the method comprises the following steps: emitting laser positioning rays and laser positioning point spots to a measured ditch and slop surface by a topographic form positioning device, wherein the laser positioning rays track the topographic form change in real time and positions the topographic form space positions, and the laser positioning points are used as control points of dimension determination and deformation correction on the topographical feature video sectional drawing; using an image data collection device for collecting the space coordinate information of point groups of the ditch and slop surface in real time; intercepting the topographical feature images before and after the time of the ditch and slop topographic form change, forming digital models before and after the time of the ditch and slop topographic form change after the processing by computer software, and finally realizing the non-contact type quantitative observation on any time of the three-dimensional ditch and slop micro topographical feature. The invention is applicable to the quantitative study on the ditch and slop micro topographical feature change process, and is particularly applicable to the quantitative study on the ditch and slop gravity etching process

(continued)

(continued)

No.	State Intellectual Property Office, China	European Patent Office (https://www.epo.org/index.html)	
	Inventor(s). Title (Grant No.)	Bibliographic data	Webpage and abstract
10	Xu X Z, Xu F L, Zhao C, Wang S F, Zhang H W. A measurement device to model the 3D landform (ZL201010144655.8)	CN101975570 (A)—2011-02-16	The invention relates to a three-dimensional observation device for scouring terrain, belonging to the field of terrain measurement and comprising a full-plane laser projector, an image collection device, a terrain image calibration device, a host lifting device, a fixed support structure and a power supply system, wherein the full-plane laser projector comprises a slot-shaped line type laser die set, a die set micro-adjusting bracket, a baffle board, a cover board, a laser distance meter and a universal joint; the terrain image calibration device comprises a spot type laser die set and a fixed supporting board; the image collection device comprises a computer, a digital camera, a camera gravity positioning box and a universal joint; the respective horizontal angle of the full-plane laser projector and the digital camera is automatically adjusted through the universal joint under the self-weight action so that the full-plane laser is in parallel to a horizontal plane; the principal optical axis of the camera is perpendicular to the horizontal plane; and the host lifting device controls the lifting of the full-plane laser projector in a step pitch setting mode. The integral set of the system has higher observation efficiency and precision, and is suitable for the research on the aspect of terrain three-dimensional analysis in a river model test
11	Xu X Z, Zhang H W, Li Z M, Zhang D C. The similarity criterions for the semi-scale experiment of soil and water conservation (ZL200610155907.0)	CN101000337 (A)—2007-07-18	A method for realizing semi-proportional model test of soil and water conservation can achieve variation of model erosion ratio by regulating antierossion ability of bedding soil itself at model so as to make sub-rainfall geomorphological evolution degree of model be nearly as the same as sub-rainfall geomorphological evolution degree of prototype

(continued)

(continued)

No.	State Intellectual Property Office, China		European Patent Office (https://www.epo.org/index.html)	
	Inventor(s). Title (Grant No.)	Bibliographic data	Webpage and abstract	
12	Xu X Z, Chai G W, Chen L. An experimental device to determine the effect of the urban porous surface collecting rainwater (ZL200810012865.4)	CN101436359 (A)—2009-05-20	The invention belongs to the technical field of research devices for water and soil conservation, and discloses a device for establishing a water-collecting effect test for urban water permeable surfaces. The device is characterized by mainly comprising a radial flow measuring device, a rainfall infiltration test surface and a rainfall simulator. The radial flow measuring device comprises a collecting groove, a collecting tank, a synchronous micro diving pump, a radial flow tank, an adjustable water control valve and a stop watch part, and adopts a transfer type measuring method to measure radial flow. The rainfall infiltration test surface uses a combined steel plate pile as a side wall to separate out an observation surface; the rainfall simulator comprises a rainfall spray head, an adjustable flat base and a water supply system; and a base bracket can keep horizontal through adjusting three lead screws respectively, so that the device is suitable for various sloping surfaces. The water supply system mainly comprises a water control valve, a water supply tank with primary and secondary structures, a first diving pump and a second diving pump. The secondary tank is positioned in the primary tank. The device has the effects and advantages of little influence on surrounding environment, accuracy and flexibility, little water usage, and low cost	
13	Xu X Z, Zhang H W, Zhang H Y. A laboratory device to simulate the process of soil and water conservation (ZL200410029760.1)	CN1563986 (A)—2005-01-12	In the invention, rainfall unit, canopy and low step face model are position at upper, middle and lower parts of bracket. Rainfall unit includes main pipe connected to water inlet pipe and branch pipes, and discharge pipe is in side opposite to the main pipe. Low step face model includes slope face, sidewall, movable lateral clapboard and movable back clapboard. Four sidewalls in model are sidewall, front wall, movable lateral clapboard and movable back clapboard. The invention includes sampler, curb and runoff bucket are arranged at proper position. Main pipe and branch pipes are made from PVC. Raininess is even. The invented equipment is suitable to indoor model test for larger area. The equipment is easy to be installed	